A Guide to Wetland Functional Design

Anne D. Marble

LEWIS PUBLISHERS
Boca Raton Ann Arbor London

This document was originally issued by the Federal Highway Administration, U.S. Department of Transportation.

Library of Congress Cataloging-in-Publication Data

Catalog information is available from the Library of Congress.

ISBN 0-87371-672-X

LEWIS PUBLISHERS, INC.
121 South Main Street, Chelsea, MI 48118

PRINTED IN THE UNITED STATES OF AMERICA
1 2 3 4 5 6 7 8 9 0

Acknowledgements

The original research and production of this book was funded by the U. S. Department of Transportation, Federal Highway Administration (FHWA-IP-090-10) in an effort to aid in the implementation of its policies.

Many individuals made the preparation of this guidebook possible. Sharon S. Yates provided the research and original draft of Chapter 11, as well as in-depth technical review of all other chapters. Katherine M. Farrow researched elements for Chapter 10, and provided extensive editing, redrafting, and formatting. Dana R. Unangst also provided essential research and internal review of this book during its preparation.

I am also indebted to numerous individuals who provided a technical review of this book in its draft form, and provided essential input. This group of peer reviewers generously gave of their own time to help make this book a useful and technically correct:

Paul Adamus, NSI, U. S. Environmental Protection Agency
Douglas Smith, WAPORA (formally of Federal Highway Administration)
Ellis Clairain, U. S. Army Corps of Engineers, Waterways Experiment Station
Michael Kaminsky, New Jersey Department of Transportation
Stuart Kehler, Pennsylvania Department of Transportation

Finally, but not least, Doris Hetrick provided the illustrations and diagrams throughout this book.

Foreword

A need exists for coherent design guidance on wetland functional replacement. It is the intent of this guidebook to provide a preliminary step in the process of fulfilling this information need.

A conceptual approach to wetland design from a functional standpoint is presented. It is based principally on the *Wetland Evaluation Technique* (WET) (FHWA-IP-88-029) which is used to determine the relative values of existing wetland functions. Site selection and site design features for wetland replacement are described for nutrient removal/transformation, sediment/toxicant retention, shoreline stabilization, floodflow alteration, ground water recharge, production export, aquatic diversity and abundance, and wetland dependent bird habitat diversity. The design of multiple functions is also discussed.

These guidelines may be used to meet wetland mitigation goals for highway projects. The guidelines should be consulted throughout the design process, beginning with the formulation of replacement goals through the final, detailed development of plans and specifications.

Table of Contents

List of Tables

List of Figures

Chapter 1
Introduction

1.1 Need for this Guidebook

Need for Functional Replacement Guidance Information. Nationally, it has been recognized that a need exists for coherent design guidance information on wetland replacement. In its final report, *Protecting America's Wetlands: An Action Agenda,* the Conservation Foundation calls for the development of basic technical guidance for functional replacement restoration design (Conservation Foundation, 1988). It is the intent of this guidebook to provide a preliminary step in the process of fulfilling this information need.

Presently, in some areas of the United States, wetland replacement is practiced on an in-kind basis. That is, the vegetation class or classes of the replacement area must be the same as those lost to a project. This approach has its obvious advantages in being straightforward, but it assumes that wetland functions correlate with vegetation structure, which may or may not in actuality be the case. In-kind replacement also may not take into consideration the importance of site location in replicating the functions of a wetland.

This guide anticipates a trend away from the approach of in-kind vegetation class replacement towards the concept of wetland functional replacement. Now that a significant amount of scientific data has been published and evaluated in relation to specific wetland functions, it is possible to begin describing wetland functional design. The information presented here describes a conceptual approach to wetland design from a functional standpoint, and has its basis in that large body of scientific literature.

Legal Basis. Over the past two decades, Federal requirements have resulted in the protection of wetlands. Examples of important Federal mandates include Executive Order 11990 - Protection of Wetlands, Section 404 of the Clean Water Act, the Coastal Zone Management Act, and the National Environmental Policy Act.

In addition, a wide range of policies and regulations have been enacted at the state level, either in the form of state environmental policy acts or as specific wetland protection acts. Wetland laws, regulations, and policies vary from state to state. Many have specific mitigation policies relating to wetland replacement, including the required replacement ratio, the location of the mitigation site (off-site vs. on-site), as well as the type of mitigation allowed (in-kind vs. out-of-kind).

Mitigation policies may also vary according to the action agency involved in a project. For example, the Federal Highway Administration (FHWA) will participate in the cost of acquiring lands sufficient to provide up to one acre of replacement for each acre of wetlands that is directly affected by a Federal-aid highway project (23 C.F.R. 777). These regulations also provide a policy regarding the location and type of mitigation measures to be considered. However, there is a considerable amount of flexibility for acceptable replacement ratios and approaches within these policies.

In summary, the specific wetland protection and mitigation laws, regulations, and policies relating to the particular project at hand must be considered and adhered to in developing wetland replacement plans.

1.2 Definitions and Assumptions

Throughout this guidebook, certain definitions of key concepts have been used. To the extent possible, the interpretation of these key concepts has been based on the most recent, widely accepted, and widely used definitions.

Classification of Wetlands. Wetlands have been classified in accordance with the U.S. Fish and Wildlife Service's *Classification of Wetlands and Deepwater Habitats* (Cowardin et al., 1979). In this hierarchical system, five principal wetland systems are defined: estuarine, lacustrine, palustrine, riverine, and marine. All but marine systems are included in this guidebook. Each of these systems are in turn divided into classes and subclasses based on dominant vegetation or substrate type.

A wetland's hydroperiod, which defines the duration and time of inundation by water, is a critical aspect of this guidebook. Definitions for the various hydroperiods are also taken directly from the *Classification of Wetlands and Deepwater Habitats,* which uses the term water regime modifier. Four tidal and seven nontidal water regimes or hydroperiods are identified.

Wetland Functions and Values. This guidebook uses the same definitions of wetland functions and wetland values as those presented in the *Wetland Evaluation Technique* (FHWA-IP-88-029) (Adamus et al., 1987). Wetland functions are the physical, chemical, and biological processes or attributes of a wetland. Values are the wetland processes or attributes that are valuable or beneficial to society.

It is assumed that the user of this guidebook has a working knowledge of wetland classification and related terminology and of the WET system of functional evaluation. This publication also assumes a basic understanding of wetland ecology and wetland definition concepts.

1.3 Basis For this Guidebook: Wetland Evaluation Technique

The informational basis for this guidebook is the *Wetland Evaluation Technique*, also known as WET. Originally developed as an aid to evaluating wetlands for highway-related projects, WET is now widely accepted and used by many agencies and organizations to evaluate existing wetland functions. However, neither WET nor any other wetland evaluation methods or models have been validated with experimental or empirical studies. WET represents an anecdotal but systematic summarization of the current knowledge of wetland functional processes.

Information from the WET manual may be used to rate a given wetland for its relative probability of performing several specific functions. Wetland functions evaluated include the following: ground water recharge, ground water discharge, floodflow alteration, shoreline stabilization, sediment/toxicant retention, nutrient removal/transformation, production export, aquatic diversity/abundance, wildlife diversity/abundance for breeding, wildlife diversity/abundance for migration and wintering, and recreation and uniqueness/heritage.

Wetland functions are evaluated using a series of predictors (questions) relating to the biological, physical, morphological, and chemical characteristics of the wetland in question. The answers to the predictors are then evaluated in separate wetland function keys. The keys evaluate the answers to the predictors and lead the evaluator to a rating of "high," "moderate," or "low" for each wetland function.

The predictors and the keys used by WET essentially describe the specific characteristics of a wetland which drive the functions and determine their effectiveness. This same information, taken apart in its component form, was used to identify specific design elements for wetland functions for this guidebook. In this way, the information contained in WET was analyzed and applied to wetland replacement design for the specific wetland functions.

Additionally, because WET is a generalized system which has nationwide application to all wetland classes, the information derived from it in turn generated generalized mitigation information for the wetland functions.

WET was used to develop design guidance by working each function key "backwards," to identify the predictors which generate a "high" rating. Only the WET predictors which contribute to a "high" rating for a particular function were evaluated for their applicability to the wetland design process. The predictors have been renamed as site selection and site design features.

The keys in WET require that certain combinations of predictors be present in a wetland to obtain a "high" rating. For design purposes, however, it was determined unnecessary to combine or group the corresponding design features into any required formula since the design features often individually contribute to a function with or without the presence of other design features. However, the relative importance of each design feature to the function under consideration is included. This is thoroughly described in Section 1.7, Guidebook Organization.

1.4 Guidebook Scope

Design Feature Emphasis. This guidebook provides conceptual design information needed to facilitate achieving functional replacement design goals for a particular wetland. The information contained herein is general enough to be easily manipulated in the site selection and site design process in a variety of physiographic settings.

This information should improve the likelihood that functional wetland replacement goals are met. A designer may either optimize for a selected wetland function or optimize for several of them given the conditions present at the replacement site.

Site Selection and Site Design. Two critical phases in the design process are focused upon: site selection and site design.

Site selection is the process by which the location of the wetland replacement site is identified. In developing a wetland design, locating the appropriate site is as essential an ingredient to the success of a wetland replacement as the site design itself. It is important to consider the physical, chemical, and biological characteristics of the site, and to determine if they will provide an adequate setting for the wetland function selected for replacement. Separate site selection features are therefore described for each of the functions.

Site design is a subsequent design phase involving the development of the actual replacement plan, that is, how the site is best manipulated to provide a given function. It includes the consideration and evaluation of design features which will optimize for the selected wetland function at the site. As with site selection, site design features are described for the separate functions.

Because of the general level of information, it will also be necessary to modify or adapt the information contained herein to relate to site specific needs. Additionally, specific design details will be required in the development of the construction plans.

1.5 Limitations

Functional Ratings. Although WET enables its user to determine the functional ratings ("high," "moderate," and "low") for existing wetlands, no similar ratings are proposed for wetland replacement until mitigation construction is complete. This is because too many unknown, site-specific factors may be present which could ultimately affect the actual functioning of the replacement wetland. Additionally, performing a WET evaluation on paper plans is risky, in that some design details affecting a wetland's functional ability could be modified in the construction phase or may differ in the field in some significant manner. It is possible, indeed encouraged, to perform a WET evaluation on the wetland after it has been constructed to verify that the desired designs have, in fact, been implemented as intended.

Although there is no guarantee that following the information in this guidebook will necessarily provide a wetland with a "high" value for the selected function, it can reasonably be assumed that the wetland will, to some degree, provide the function. Generally, the more design features (particularly those with a high degree of importance) which are included in a plan for a given function, the higher its potential value.

Self-sustainability. This guidebook assumes that the primary goal of wetland replacement is to optimize for a selected function or functions. However, because wetlands are dynamic systems, a particular function may not be sustained in a given wetland over time without periodic long term maintenance. In these situations, functional optimization may not be compatible with self-sustainability, and a choice between these two goals will have to be made.

Level of Design Detail. This guidebook is not intended to be a procedure manual. Rather, it provides a conceptual starting point for the implementation of wetland functional design features into a replacement plan. The development of detailed wetland replacement plans, costs, and construction drawings are not described.

Use of Specific Criteria. Specific criteria (e.g., specific distances, percentages, etc.) should not be used as absolute limits when employed in site design. The criteria are based on the original predictors presented in WET, which are used to obtain ratings for specific functions in that method. In most cases, these criteria were based on data reported in the literature (which may not apply to all wetland types) or were empirically derived.

When criteria are indicated in this guidebook, they are only suggested limits. For example, it is suggested that a vegetated width of 500 feet or greater be used to maximize for nutrient removal/transformation. This does not suggest that a width of less than 500 feet of vegetation will not perform this function. Generally, 500 feet in this example is a somewhat arbitrary number, but one which has been supported by studies of certain wetland types and can be used as a design goal. If possible, criteria should be adapted to the region in which the project takes place.

Confidence Level. For each function, several design features are recommended for inclusion in the site selection and site design process. It is important for the user to consider that much of this information has not been previously used or evaluated in the field. Rather, these design recommendations often arise directly from WET as an evaluation tool.

Wetland Systems Included. Estuarine, palustrine, riverine, and lacustrine systems are included, but marine systems are not. In the majority of cases involving marine systems, wetland replacement is not a recommended mitigation strategy.

1.6 Wetland Replacement Design in the Impact Evaluation/ Mitigation Process

Replacement design represents one of many steps in the regulated wetland impact and mitigation process. Before consulting the information contained herein, it is assumed that several previous activities and agreements have taken place. Although the order of these activities and specifics of the procedures vary from state to state, they are summarized in a general manner below:

- All wetlands have been delineated in accordance with the *Federal Manual for Identifying and Delineating Jurisdictional Wetlands* (Federal Interagency Committee for Wetland Delineation, 1989), and the limits of the boundaries have been agreed upon with appropriate regulatory agencies, including the U.S. Army Corps of Engineers.

- An alternative analysis has been conducted in accordance with requirements set forth in Section 404 (b)(1) of the Clean Water Act and 23 C.F.R. 771 of the Federal Highway Administration. The results of this analysis indicate that all efforts have been made to avoid and minimize harm to wetlands involved with the project.

- Wetland restoration or creation has been determined to be the accepted mitigation strategy.

- A determination has been made regarding the exact number of acres requiring replacement as a result of the project, and the replacement acreage has been agreed upon with appropriate regulatory agencies. The acreage of the wetland replacement site may also be determined based on the function to be replaced, using this guidebook to assist in developing replacement ratios.

- A WET evaluation or similar accepted functional assessment has been conducted on the existing affected wetlands to determine the rating of each function.

- A general understanding has been reached with all involved regulatory, resource, and action (originating) agencies regarding the wetland functional replacement design goals for the project. The need for continued coordination with involved agencies has also been established.

This guidebook may be consulted throughout the design process beginning with the formulation of replacement goals, through the final, detailed development of plans and specifications. At the beginning of the design process the user is guided to consider site characteristics affecting the selected wetland function. In the site design phase, the physical and biological design components which can be manipulated to optimize for the selected wetland function are identified. It is essential that additional, detailed site-specific and project-specific data and information be collected and evaluated during the design process and during the follow-up monitoring period.

The final stage of the replacement design process involves the preparation of detailed site grading and planting plans and specifications. This information, along with lists of quantities of materials, costs, and site construction provisions, must be included in a final design package for use by the construction contractor. The requirements of the State highway agency will provide the guidance necessary for preparing the final plans, and should be consulted accordingly.

Role of Regulatory, Resource, and Action Agencies. When wetland replacement is determined to be the accepted mitigation strategy for a project, it is essential that the appropriate regulatory and resource agencies be involved in the replacement plan development. Early and continuous involvement of all interested parties will build the specific wetlands expertise of each agency representative into the design process. It will also provide a forum to decide and agree upon specific replacement.

The type and degree of agency involvement depends on the established review procedures, but activities may include the following: the selection of the function or functions to be replaced by the wetland as well as the identification of additional, more specific goals; the selection of the type and location of the replacement site; the development and/or review of design concepts; and the review of final design construction drawings and specifications. Additional agency involvement may be necessary or useful depending on the specific project under consideration.

1.7 Guidebook Organization

This section describes the step-by-step procedure for using this guidebook. It also describes its format and layout.

Wetland Functions Included. Generally, the wetland functions described follow those of WET. Sediment Stabilization has been renamed in this guidebook as Shoreline Stabilization. Wildlife diversity and abundance for breeding, migration, and wintering are combined. Because the WET assessment for wildlife diversity and abundance is principally for wetland dependent birds, this function has been renamed Wetland Dependent Bird Habitat Diversity.

The following functions are described on a chapter by chapter basis in this guidebook:

Chapter 3.	Nutrient Removal/Transformation
Chapter 4.	Sediment/Toxicant Retention
Chapter 5.	Shoreline Stabilization
Chapter 6.	Floodflow Alteration
Chapter 7.	Ground Water Recharge
Chapter 8.	Production Export
Chapter 9.	Aquatic Diversity/Abundance
Chapter 10.	Wetland Dependent Bird Habitat Diversity

Wetland Functions Not Included. The functions of ground water discharge, recreation, and uniqueness/heritage were not included. It was considered unlikely that wetland replacement goals would include ground water discharge as a function, since ground water discharge is often not a direct attribute or function of the wetland itself. Rather, it is an aspect of wetland hydrology. Recreation and uniqueness/heritage were not included because no scientific basis exists for an objective assessment without site specific and project specific considerations. This function of wetlands is important, however, and can in fact be considered as a replacement goal.

Format of Wetland Function Chapters. The format of the wetland function chapters is similar. Each chapter includes a brief description of the function and general design concepts, followed by a matrix summarizing design features. A design feature is a variable which can be manipulated in the design process to optimize for the selected function.

The matrix also categorizes each design feature according to its applicability to site selection and site design. Those features described for site selection may be considered in the process of locating an appropriate wetland replacement site. Site design features may be considered in the development of the wetland replacement plan. Some features are applicable to site selection as well as site design, and therefore should be considered in both processes.

The summary matrix also rates each design feature "high," "moderate," or "low" depending on its relative importance to the function. Those features having a high importance rating should be considered and included first, followed by those rated moderate and then low. It should be noted that these ratings of "high," "moderate," and "low" do not relate to the individual function ratings derived from a WET assessment.

Another purpose of the importance rating is to give the user a basis to make design feature tradeoffs. For example, if a decision must be made to include one design feature over another in a plan, the feature with a higher relative value may be selected. A tradeoff may be necessary if two design features conflict with each other, or if the physical properties of the replacement site allow only one of the two to be included.

In general, however, the goal of the wetland replacement design is to incorporate as many design features as possible into the site selection and site design process, since they all contribute to some degree to increasing the wetland's value for that function.

Each design feature listed in the matrix is identified by a section number. This section number serves as a reference for subsequent, detailed descriptions for each design feature section in that chapter.

The summary matrix for each function also identifies those WET predictors which were not used. These predictors were either not applicable to the design process, or were described more succinctly by other predictors.

Design Feature Sections. The individual Design Feature Sections are the heart of this guidebook. For each section, the following information is supplied:

Type: Site selection or site design feature. This also refers to the stage in the process to which the design feature should be applied. In some cases, the feature may be applied to both site selection and design.

System: The wetland system to which the feature may be applied.

Criterion: Definition of the feature.

Rationale: Why the feature is important to the wetland function.

Methods: How to incorporate the design feature into the site selection or site design process. In some cases where methods are similar, the user may be referred to another chapter in the guidebook.

Notes: Important exceptions and information relating to the design feature.

Hydrology: the Driving Force. In addition to the above chapters relating to the specific wetland functions, a separate chapter on wetland hydrology is included (Chapter 2). Hydrology, and more specifically hydroperiod, is a critical, basic element in the functional ability of a wetland. As such, it plays a fundamental role in determining the wetland's relative value. Because many other design features often depend on the hydroperiod and because hydroperiod ultimately determines to a large degree the relative success of a wetland replacement, it is important that this design feature be developed as accurately and fully as possible.

For many wetland functions, a specific type of hydroperiod or hydrologic regime is designated. Each wetland function chapter (Chapters 3 through 10) describes the hydroperiods recommended specifically for that function, and in turn refers the user to Chapter 2. Chapter 2 describes the methods for deriving each hydroperiod on a site-specific basis. Data collection and monitoring requirements are described for surface water and ground water sources, and for tidal and nontidal systems. In summary, all hydroperiod information is summarized in Chapter 2 for a cohesive discussion and comparison.

1.8 Designing for Multiple Function Replacement

Wetland mitigation plans can often be developed for the purpose of replacing several wetland functions. Chapter 11 describes the process of designing for multiple functions. Areas discussed include compatibility of functions, compatibility of site selection and site design features within a function, and compatibility between multiple functions.

1.9 Relationship to Other Mitigation Reports

Other efforts on wetland mitigation have been completed or are in progress. These studies are useful to the readers who may wish to expand their level of understanding and detail in relation to wetlands creation, or to develop it more fully for specific functions.

A partial list of other wetland mitigation reports includes the following:

- *Criteria for Created or Restored Wetlands* (Adamus, 1988).
- *Management of Artificial Lakes and Ponds* (Bennett, 1971).
- *Habitat Management Models for Selected Wildlife Management Practices in the Northern Great Plains* (Sousa, 1987).
- *Wetland Creation and Restoration: The Status of Science* Vol. 1 and Vol. 2 (Kusler and Kentula, 1989).
- *Management of Seasonally Flooded Impoundments for Wildlife* (Fredrickson and Taylor, 1982).
- *Wetland and Riparian Habitats: A Nongame Management Overview* (Fredrickson and Reid, 1986).
- *Fish Habitat Improvement Handbook* (U.S.D.A. Forest Service, 1985).
- *Instream Habitat Improvement Devices and Their Uses in Freshwater Fisheries Management* (Swales and O'Hara, 1980).
- *Guidelines for Management of Trout Stream Habitat in Wisconsin* (Wisconsin Department of Natural Resources, 1967).

Chapter 2
Wetland Hydrology

By definition, all wetlands are created and maintained by water. The frequency, depth, and duration of the water's influence determine, to a significant extent, the vegetation present and the functions that the wetland provides. Water, whether from a surface water source or from ground water, is the most critical feature to define and evaluate in attempting to reproduce a naturally occurring wetland system.

In order to create a wetland system which provides specific functions, one specific hydroperiod or range of hydroperiods is often most effective or desirable. In the following chapters, specific goals in site selection and site design are discussed in relation to specific wetland functions. For many functions considered, hydroperiod is a critical feature. The following information is provided as guidance in designing for a specific wetland hydroperiod once the water source and desired hydroperiod have been determined.

The specific hydroperiods required for each function are described in the corresponding function chapters (Chapters 3 through 10). Each of these chapters refers to the methods described below for creating the specific hydroperiod required. The methods presented herein are intended to be general guidelines. Regional and site specific information should be obtained in order to design a specific wetland hydroperiod.

2.1 Hydroperiod Definitions

Hydroperiod is defined as the periodic or regular occurrence of flooding and/or saturated soil conditions. Wetland hydroperiods are defined and classified according to the water regime modifiers described in the U.S. Fish and Wildlife Service publication entitled *Classification of Wetlands and Deepwater Habitats of the United States* (Cowardin et al., 1979). The following are definitions of 12 hydroperiods which occur in wetlands in accordance with this system.

Tidal Hydroperiods. There are four tidal hydroperiods defined by the U.S. Fish and Wildlife Service. The tidal hydroperiods are illustrated in Figure 1, and defined below:

- *Irregularly Flooded.* Tidal water floods the land surface less often than daily.
- *Regularly Flooded.* Tidal water alternately floods and exposes the land surface at least once daily.
- *Irregularly Exposed.* The land surface is exposed by tides less often than daily.
- *Subtidal.* The substrate is permanently flooded with tidal water.

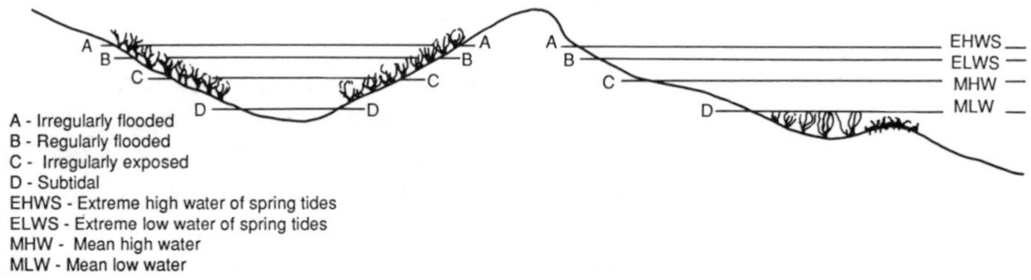

Figure 1. Tidal hydroperiods (Source: Cowardin et al., 1979).

Of the tidal hydroperiods, subtidal has soils which are permanently saturated while the irregularly flooded has soils which are infrequently saturated.

Nontidal Hydroperiods. The eight nontidal hydroperiods, with the exception of intermittently flooded and artificially flooded, are illustrated in Figure 2.

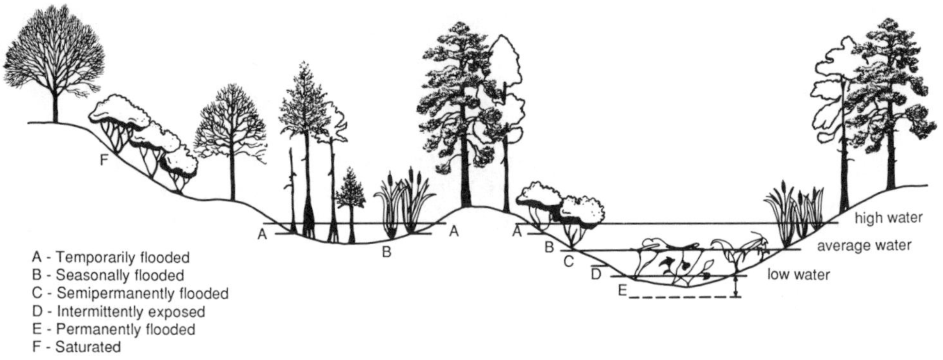

Figure 2. Nontidal hydroperiods (Source: Cowardin et al., 1979).

The nontidal hydroperiods are defined below:

- *Temporarily Flooded.* Surface water is present for brief periods during the growing season, but the water table usually lies well below the soil surface for most of the season.

- *Seasonally Flooded.* Surface water is present for extended periods especially early in the growing season, but is absent by the end of the season in most years. When surface water is absent, the water table is often near the land surface.

- *Semipermanently Flooded.* Surface water persists throughout the growing season in most years. When surface water is absent, the water table is usually at or very near the land surface.

- *Intermittently Exposed.* Surface water is present throughout the year except in years of extreme drought.

- *Permanently Flooded.* Water covers the land surface throughout the year in all years.

- *Saturated.* The substrate is saturated to the surface for extended periods during the growing season, but surface water is seldom present.

- *Intermittently Flooded.* The substrate is usually exposed, but surface water is present for variable periods without detectable seasonal periodicity (not illustrated).

- *Artificially Flooded.* The amount and duration of flooding is controlled by means of pumps or siphons in combination with dikes or dams (not illustrated).

2.2 Designing for Specific Hydroperiods

Tidal Systems

After selection of a replacement site, locate the existing and proposed mean tide elevations during the growing season. The principal elevations to define include mean high water (MHW) and mean low water (MLW) as well as the spring tides (EHWS and ELWS).

The existing mean high and mean low water elevations for the replacement site should be located at and immediately adjacent to the replacement site. These elevations should be tied into a topographic and/or bathymetric (bottom topography) survey (Figure 3). The existing topographic survey should show contour intervals of no more than one foot.

This information is used to develop a grading and planting plan for the replacement site. Contour intervals for these plans should be at least one foot or six inches, depending on the types of plants selected for use and/or the tidal range of the wetland.

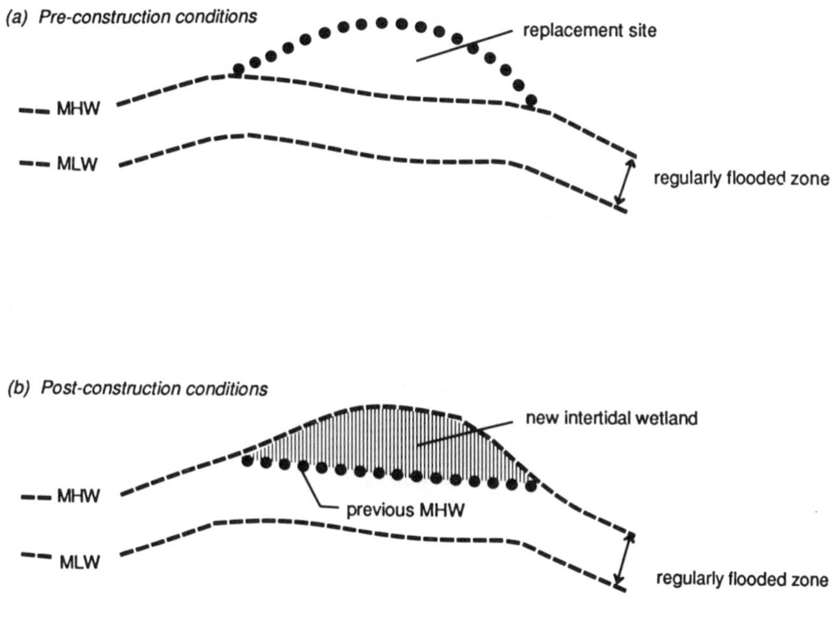

Note: This type of plan must include an evaluation of long term accretion and erosion patterns which could eventually alter the site and affect the wetland's function.

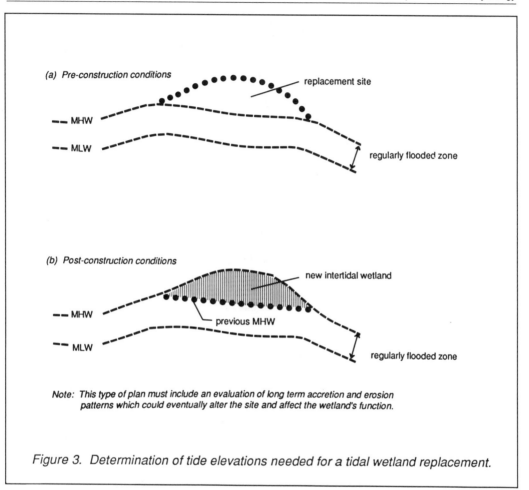

Figure 3. Determination of tide elevations needed for a tidal wetland replacement.

Because tidal range decreases to the south, a one foot contour interval may be too coarse for some southern states.

Existing wetlands adjacent to the replacement site may also be correlated to existing topographic elevations. The topographic zones inhabited by each community type of the existing wetland should subsequently be determined. Use these data as a reference for the replacement site's vegetation communities.

Field data should be compared to published data on tide elevations and/or locations. Tide information obtained from the

National Oceanic and Atmospheric Administration and other agencies should be carefully reviewed to make sure elevations are based on the same datum as the grading plans for the wetland replacement. Also, it should be determined if there are any tide control structures that could affect the elevations at the replacement site.

Nontidal Systems

Wetlands with a Ground Water Source. The general elevation of the water table at a replacement site can be determined preliminarily from existing data. This information should not replace field data. Data to collect includes:

- *Soil Survey.* Information for specific soil series can be obtained from the county soil survey maps. The description of the soil series at the site will indicate the depth of the seasonal high water table during average conditions. Also, depth to bedrock and soil permeability rates can be obtained from soil surveys.

- *Soil Boring Logs.* If available, soil boring logs for adjacent highway projects taken nearby can be interpolated for the replacement site. This is useful only if the site is immediately adjacent to the right-of-way and if the underlying soils and geologic conditions are similar.

- *Location of Adjacent Wetlands.* The elevation of nearby existing wetlands with a ground water source is often useful. This assumes that the water table for the replacement site is the same as for the adjacent site and, therefore, must be used with some caution. Check to see if the underlying soils and geologic conditions are similar.

- *Streambed Elevation.* If present, nearby perennial streams may provide an indication of the adjacent ground water table elevation. Stream profile elevations can be tied in to other sources of data to establish the ground water contours throughout the site.

After collecting and evaluating preliminary data, it is highly advised that a well monitoring program be established at the replacement site. The information collected during this program is critical in establishing the necessary hydrologic conditions for the wetland. Place a series of monitoring wells at representative locations throughout the site. The preliminary data will provide guidance on the depth to place the wells, the frequency of readings, and the appropriate seasons to conduct monitoring.

Figure 4 depicts a cross section of a typical monitoring well for ground water. Use a 0.75 to 1.0 inch diameter PVC pipe as the monitoring well. A slotted bottom section will enable the movement of water from the soil into the well. Cap and vent both ends of the pipe. Backfill the boring hole with fine grained gravel or coarse sand. Use bentonite and concrete to seal the

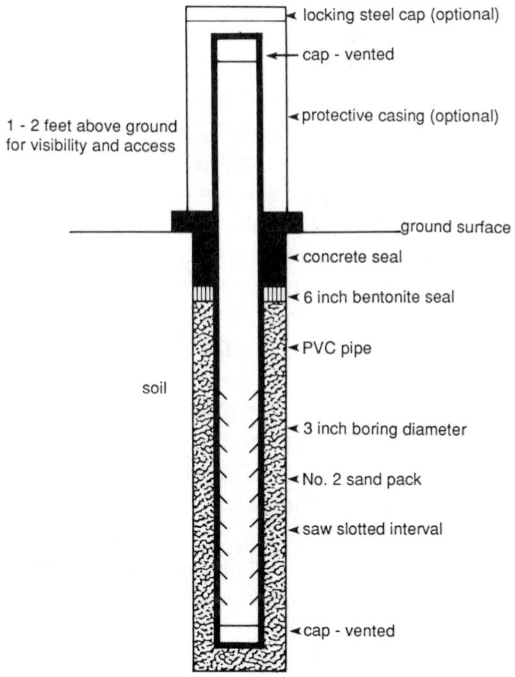

Figure 4. Monitoring well schematic cross section.

upper portion of the well from surface water inflow. A steel locking cap may be used for well protection during the monitoring period.

Some states may have specific design requirements for monitoring wells. This should be determined in advance of developing details for the monitoring plan. Also, permits may be needed to install monitoring wells.

Prior to placement of the wells, investigate the soil profile to determine the appropriate depth for each well. The county soil survey will provide some general information as to the depth to the water table for each soil series. Be sure to place the wells deep enough to intersect the local water table, not a seasonally perched water table, unless a perched water table is specifically desired as the water source for the wetland.

Read water levels in the wells at regular intervals throughout the monitoring period. Correlate the data to on-site precipitation data, and in turn, compare it to the average recorded rainfall data for the vicinity. This will determine if the rainfall, and hence the monitored water table level at the site, is at a normal elevation for that time of year. If rainfall data for the monitored year is only slightly higher or lower than normal, the monitored levels may be used to determine the wetland water elevations. If the rainfall data indicates that the season is significantly wetter or drier than normal, it may be necessary to continue well monitoring for a longer period of time.

Conduct ground water monitoring from the period of seasonal high water through the period of seasonal low water during the growing season. Data from several wells which are placed on a transect perpendicular to the slope can be used to plot the elevation of the water table across the site. Once the water table elevation is known, determine the depth of excavation required to create a specific hydroperiod as follows:

> To create *saturated soil conditions,* excavate to the elevation of the seasonal high water table. The seasonal high water table must be maintained for a minimum of 30 consecutive days.

- To create a *seasonally flooded hydroperiod*, excavate the site to an elevation between the seasonal high and normal water table level. This will allow the water level in the wetland to fluctuate seasonally, with the soil being inundated at times and saturated at others.

- *Semipermanently flooded conditions* require excavation of the site to the depth of the seasonal low water table depth. The wetland will remain flooded throughout the growing season in most years and the water table occasionally will drop to a level slightly below the ground surface.

- *Intermittently exposed conditions* require excavation to a depth below the seasonal low water level. Flood conditions will persist throughout the year with the loss of water only in years of drought.

- *Permanently flooded conditions* may be created in the same manner as intermittently exposed conditions. However, it may be difficult to ensure the presence of water during drought years unless historic records indicate a stable water level during previously recorded drought years. These types of records are rare.

If the ground water table as monitored does not fluctuate seasonally it may be difficult to design for some hydroperiods, such as seasonally flooded and semipermanently flooded. In this case it may be desirable to include a surface water source to provide some seasonal fluctuation in the amount of water available in the wetland.

Wetlands with a Surface Water Source

Two principal surface water sources may be used: runoff or an existing perennial (permanent) stream. By definition, a perennial stream also includes runoff, but the design approach to the two types of water sources is different.

It is important to note that certain hydroperiods cannot be designed using a surface water source of either type. These include semipermanently flooded and saturated hydroperiods, which must depend on a ground water source. All other hydroperiods may be designed using a surface water source only.

If a perennial stream is used as the water source, the following guidelines may be used for specific hydroperiods.

For *intermittently exposed* and *permanently flooded* hydroperiods, use an embankment type basin, building a retention basin in conjunction with a stream. A variation of this is an embankment basin which is also excavated (see Section 2.3). Design the spillway or outlet of the basin at an elevation which will maintain water throughout the year. Other factors will need to be considered, including the off-site effects on flooding, safety, fish migration, and downstream uses of the stream. Permits may be necessary to obtain for a waterway modification.

If a lacustrine site is used, ensure that the seasonal low water level of the lake does not drop below that of the replacement wetland. If conditions are unknown, it has been shown that monitoring the water level conditions at the time of seasonal low water will provide information necessary to ensure an adequate water supply for the wetland. The type of water control structures present on the lake, if any, should also be checked for their impact on water levels.

For *temporarily flooded, seasonally flooded,* and *intermittently flooded* hydroperiods, the wetland basin may be placed immediately adjacent to an existing stream. The cross sectional area of the existing stream can be modified to include a wetland basin, as shown in Figure 5. A hydraulic study will be necessary to determine the cross sectional areas of the stream, flood volumes, and the elevation of the flooded area for pre-construction and post-construction conditions. The cross section and flood elevation data are used to determine the topographic modifications necessary to create wetland conditions. The desired flood duration depends on the hydroperiod selected.

If surface water in the form of stormwater runoff is used, the amount of runoff generated from a watershed may be estimated using Soil Conservation Service's Technical Release No. 55, *Urban Hydrology for Small Watersheds* (1975). This method estimates runoff quantities using volume parameters such as soil and cover type and time parameters such as slope, flow

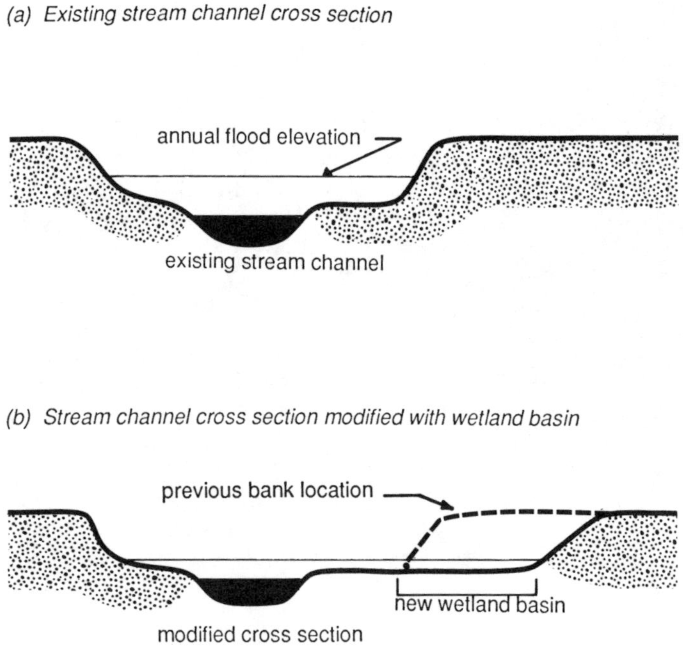

Figure 5. Using an existing stream channel as a wetland basin.

length, and surface roughness. Figure 6 can be used to estimate the minimum size drainage area required for each acre-foot of storage in a wetland using runoff as its main source. Figure 6 is intended for use as a general guide. If reliable local runoff information is available, it should be used in place of Figure 6.

In some regions, loss of water by evaporation, transpiration, and seepage are too great to enable the establishment of wetland conditions if surface runoff is the only water source. These cases will require the use of a permanent water supply in addition to surface water runoff. The wetland can be

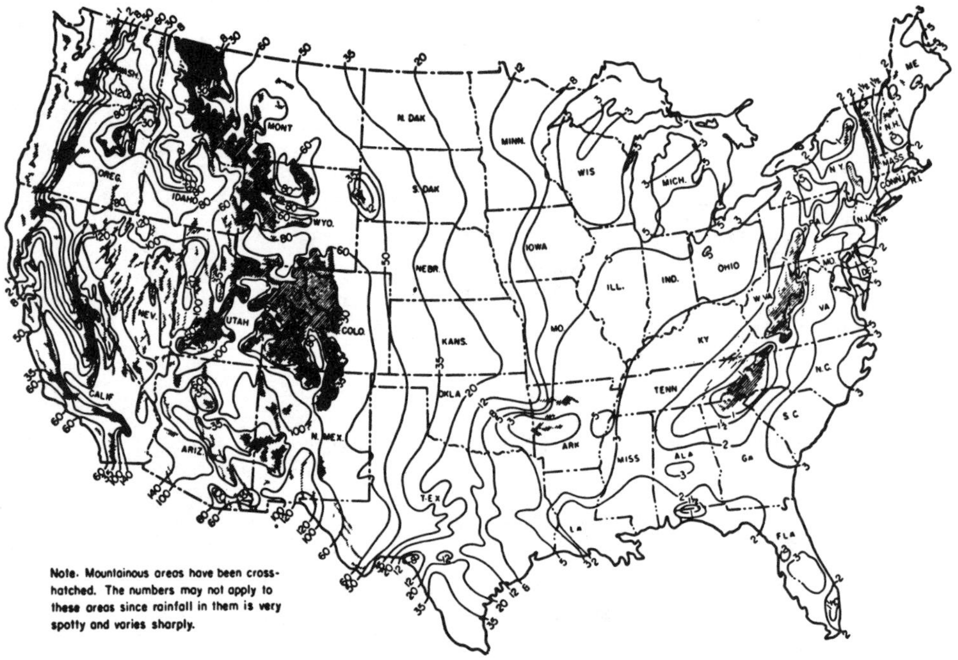
Note. Mountainous areas have been cross-hatched. The numbers may not apply to these areas since rainfall in them is very spotty and varies sharply.

Figure 6. A guide for estimating the approximate size of drainage area (in acres) required for each acre-foot of storage in an embankment or excavated pond
(Source: U.S.D.A. Soil Conservation Service, 1971).

designed to take advantage of permanent stream flows during periods of low or sporadic rainfall.

Deepwater habitats including ponds and lakes exceeding 5 to 10 feet in water depth may be designed to take advantage of stormwater runoff as the primary water source. The large storage capacity and morphology of these basins allow for storage of water over longer periods. Although the fringes may experience a significant drop in the water level late in the growing season, or during drought periods, the deepwater habitat will be maintained.

Several sources are available which detail how to compute capacities for permanent water storage. These publications are based on lake and pond construction but can be adapted for wetlands using surface water as a water source:

- *Ponds: Planning, Design and Construction.* (U.S.D.A. Soil Conservation Service, 1971c).
- *Ponds for Water Supply and Recreation.* (U.S.D.A. Soil Conservation Service, 1971b).
- *Building a Pond.* (U.S.D.A. Soil Conservation Service, 1971a).
- *Real Estate Lakes 601-G.* (Geological Survey Circular, 1971).

The local Soil Conservation Service office should be contacted for data related to pond construction for specific regions. The amount of runoff available during the growing season and runoff generated by storm events should be considered in locating and sizing the wetland basin.

The potential for seepage losses can be estimated by determining the permeability of the soil or substrate to be used and the ground water gradient. If either seepage or soil infiltration losses are high due to a porous soil substrate, the basin may be sealed with a compaction seal, a less permeable soil type, or an artificial substrate.

Losses of water due to evapotranspiration should be considered in subhumid and semi-arid climates. The Blaney-Criddle procedure is sometimes used to estimate these losses (Blaney, 1961 and SCS, 1970).

2.3 Nontidal Wetland Basin Design Types: A Summary

Two principal types of basins may be used for nontidal wetland replacement, depending on the water source (Figure 7). An excavated basin is dug below the original ground surface elevation. This basin type is used for wetlands whose principal water supply source is ground water, although it is also

(a) Excavated basin

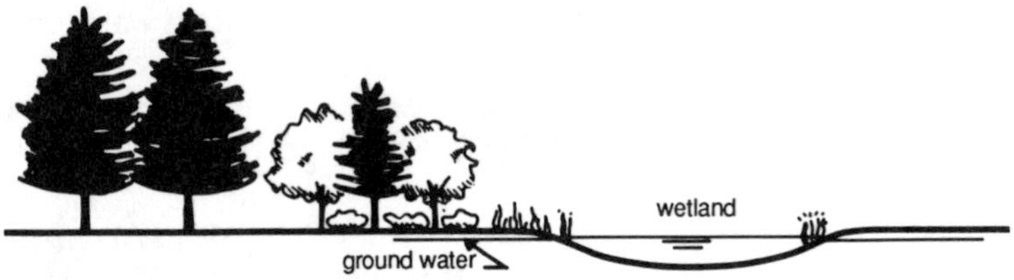

Excavated basins are dug below the original ground elevation.

(b) Embankment basin

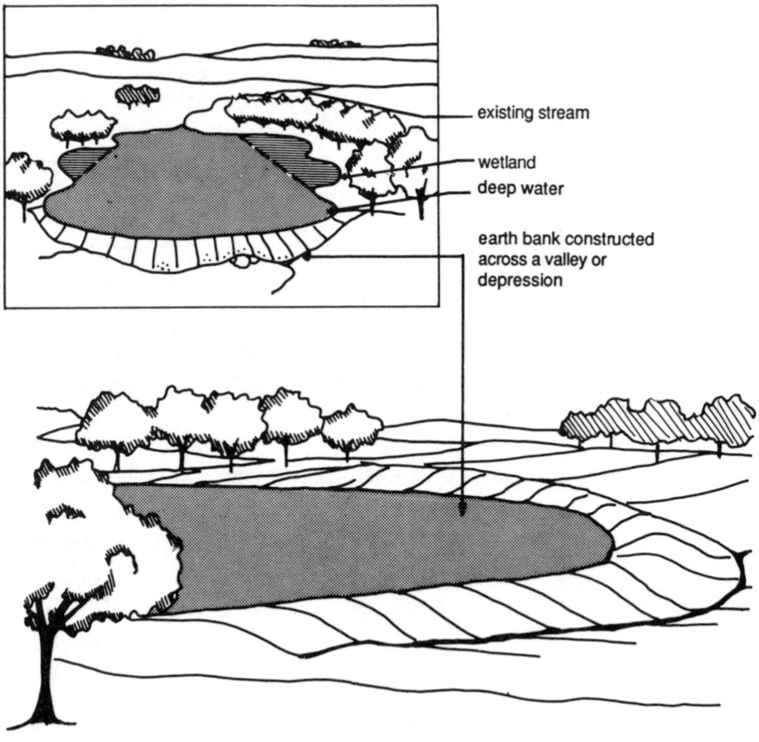

Figure 7. Wetland basin types (Source: U.S.D.A. Soil Conservation Service, 1982).

useful for surface water source wetlands or a combination of both.

The embankment type of basin is used only for wetlands with a surface water supply. It consists of an earthen berm constructed across an existing valley or depression. A combination of the embankment and excavated basin is also feasible.

2.4 Sedimentation and Erosion Considerations

For wetlands using a surface water source, the sediment load of the incoming water and outgoing water should be considered. If incoming sediment loads are high, the basin's storage capacity may be compromised over time, ultimately affecting the wetland's ability to function.

In other situations, a wetland acting as a sediment trap may so significantly alter the sediment load in a stream as to cause downstream scouring and erosion. In these cases, downstream floodplain sediments may be resuspended. A detailed evaluation of these conditions should be made with the assistance of a professional sedimentologist and/or sedimentation models.

Chapter 3
Nutrient Removal/Transformation

Description of the Function:

This function involves the retention of nutrients, the transformation of inorganic nutrients to their organic forms, and the transformation of nitrogen into its gaseous form. Because nitrogen and phosphorus are usually of great importance to wetland systems, emphasis is placed on these nutrients.

Wetlands maintain water quality of receiving waters by removing nutrients. Nutrients can be removed from both the water column and sediments during the growing season.

On a short term basis, nutrients can be taken up and stored by wetland vegetation. Once the plants die or defoliate, the nutrients are returned to the water and sediment. These nutrients are released when light or temperature is less likely to support excessive algae growth. On a long term basis, wetland vegetation may remove nutrients from biological cycling through sedimentation by aiding burial below the zone of biological activity. Nutrients may be removed by physically burying the sediments to which they are attached. Sediment particles are removed by dense wetland vegetation which slows the flow of sediment laden water. The slower the water velocity, the greater the settling of sediments from the water column and thus the increase of nutrient burial.

Several chemical and microbial processes also function to remove or transform nutrients. These processes are described below.

Denitrification. This process results in the permanent loss of nitrogen from the wetland. Denitrification is the conversion of dissolved nitrogen to gaseous nitrogen by microbes in anaerobic conditions. Because it is dependent upon nitrate formed under aerobic conditions, this process proceeds most rapidly with fluctuations in or in close proximity to oxygenated and anoxic conditions.

Nitrogen Fixation. This process is the opposite of denitrification, and may be the source of significant nitrogen for some wetlands. It involves the conversion or fixation of gaseous nitrogen into inorganic forms by bacteria and blue-green algae.

Ammonium Volatilization. This process also involves the removal of nitrogen from wetlands. Ammonium volatilization is an abiotic process which results in the removal of ammonium by evaporation. The process occurs at high temperatures and at a pH of greater than 7.5.

A final, relatively minor mechanism for nitrogen removal is biotic in nature. Seasonal and sometimes permanent losses of nitrogen from wetlands may result from the seasonal emergence of aquatic insects and consumption of nutrient rich aquatic plants by migrating waterfowl.

Phosphorus interactions occur within wetland sediments. Phosphorus is readily immobilized by calcium, aluminum, and iron by adsorption and precipitation reactions. Because fine mineral soils usually have higher concentrations of these ions, they typically have higher capacities to retain phosphorus than organic soils.

Generally, studies show that many wetlands act as sinks for nitrogen and phosphorus under nutrient enriched and natural conditions. Therefore, wetlands can function as nutrient traps by intercepting urban or agricultural runoff containing high concentrations of dissolved nitrogen and phosphorus before they reach lakes, rivers, and reservoirs. The efficiency of nutrient removal and transformation is greater with longer retention times and relatively low loading rates.

General Design Concepts:

A wetland whose primary function is to retain and transform nutrients must primarily be capable of physically detaining the nutrients. This is accomplished when the water velocity entering a wetland slows so that sediments and their adsorbed nutrients may settle to the bottom of the basin. Low gradients encourage low velocity which in turn allow for sedimentation (Section 3.5).

A related physical characteristic is long term detention of surface water entering the wetland. Water detained for a long period of time has greater potential for sedimentation and biological processing of nutrients in the water column (Section 3.6), but biological processes may be limited by the oxygen content of the water.

Wetland plants function in several ways to remove nitrogen and phosphorous on a short term and long term basis. In particular, the stems and leaves of emergent and multi-stemmed woody plants offer frictional resistance to incoming surface water and enhance nutrient retention by burial.

Wetland vegetation also acts to remove nutrients seasonally from the sediment and water column by biological growth processes. Emergent and aquatic bed vegetation are particularly useful in nutrient uptake, although woody vegetation can store nutrients for a longer period in their woody tissues (Section 3.7). Other factors being equal, the more dense the vegetation is, the greater the wetland's ability to remove and take up nutrients (Section 3.8).

Soil conditions also determine the degree to which a wetland removes nutrients. It is known that permanently saturated or flooded wetland soils favor phosphorous retention. Phosphorous is also most easily removed by complexing with aluminum, calcium, and iron in wetland sediments. The close association of anaerobic and aerobic conditions at the surface of saturated sediments, and the rapid fluctuation between anaerobic and aerobic conditions favor nitrogen removal (Section 3.4).

Specific Site Selection and Site Design Features:

The following table describes the features used in the nutrient removal/transformation function. With the exception of Section 3.2, Water Source, these features were derived from the WET predictors. The site selection and site design features may be applied to all wetland systems.

Table 1. Nutrient removal/transformation site selection and site design features.

Section	Feature [1]	Site Selection	Site Design	Importance to Function [2]	Notes [3]
3.1	Nutrient Sources	X		Moderate	26
3.2	Water Source	X	X	High	new
3.3	Hydroperiod		X	Moderate	33
3.4	Soils and Water Alkalinity	X	X	High	24,56
3.5	Channel Gradient and Water Velocity		X	Moderate	7,41
3.6	Outlet Characteristics		X	High	8,9 renamed
3.7	Vegetation Class and Form Richness		X	Moderate	12,17
3.8	Vegetated Width		X	Moderate	36
3.9	Wetland/Watershed Ratio	X		Moderate	5

[1] Each feature applies to estuarine, riverine, lacustrine, and palustrine wetland systems.

[2] These ratings are generally derived from Volume I of WET (FHWA-IP-88-029). Some values were modified in relation to wetlands replacement criteria by Paul Adamus (U.S. EPA Environmental Research Lab, Corvallis, Oregon).

[3] This column describes the WET 2.0 predictors used in deriving each feature. If only a number is given, it refers to the predictor number in WET 2.0. RENAMED means that the title of the WET 2.0 predictor was modified to show the new feature. NEW means the feature is not based on any given WET 2.0 predictor, but was inferred from other information.

WET 2.0 Predictors Not Used

1 Climate; 4 Location and Size; 21 Land Cover of the Watershed; 23 Ditches/Canals/Channelizations/ Levees; 28 Direct Alteration; 56 Dissolved Solids or Alkalinity

Section 3.1
Nutrient Sources

Type: Site selection

System: Estuarine, riverine, lacustrine, palustrine

Criterion: Locate the replacement site in a watershed where it will be exposed to moderate loadings of incoming nutrients.

Rationale: The presence of nutrient rich sources of surface water provides a high opportunity for a wetland replacement area to remove and transform nutrients. Constricted wetlands are commonly proposed for use as nonpoint treatment facilities.

Methods: Locate the replacement site in a watershed containing one or more sources of nutrients such as:

- sewage outfall,
- phosphate mine,
- canals,
- feed lots,
- pasture land,
- landfills, and
- urban runoff.

Loading rates of incoming water should not exceed the wetland's ability to assimilate nutrients, and have the potential to adversely affect plant growth, fish, wildlife, or safety. Therefore, it is recommended that only moderate loadings of nutrients be present in the incoming water.

Section 3.2
Water Source

Type: Site selection and site design

System: Estuarine, riverine, lacustrine, palustrine

Criterion: Use a surface water source as the wetland's principal water supply.

Rationale: Water supply originating from a surface water source, such as channel flow, overland flow, and precipitation, will carry nutrients into the wetland from the watershed. In comparison, ground water sources are less likely to contain high levels of nutrients and thus provide less chance to perform this function. The wetland may therefore have any type of inlet (none, intermittent, or permanent) as long as it is receiving surface water inflow.

Methods: Generally, avoid using excavated basins which intersect the water table as a principal water supply source for the wetland. Excavated ponds can be used to contain surface runoff, but well monitoring to the depth of the proposed basin should take place to ensure that ground water will not be the major input into the basin. Generally, it is recommended that ground water not be allowed to enter an excavated pond.

Impoundment or embankment basins, which consist of a dam or bank of earth constructed across a valley or depression, use only surface water as a water supply source. These types of basins store water primarily above the natural ground level.

Section 3.3
Hydroperiod

Type: Site design

System: Estuarine, riverine, lacustrine, palustrine

Criterion: For nontidal areas, create permanently flooded or permanently saturated conditions. For tidal areas, irregularly flooded conditions are most favorable.

Rationale: Nitrogen is best removed by permanently flooded or saturated conditions in nontidal areas, or by irregularly fluctuating conditions found in tidal and floodplain areas.

Methods:

Nontidal Systems
Create a wetland whose water regime is permanently saturated or permanently flooded. In permanently flooded conditions, water covers the land surface throughout the year. Saturated conditions are where the substrate is saturated for extended periods during the growing season.

Tidal Systems
Create a wetland whose water regime is irregularly flooded or irregularly exposed. Regularly exposed regimes are also desirable but to a lesser extent.

Methods for creating specific hydroperiods are discussed in Chapter 2, Wetland Hydrology.

Section 3.4
Soils and Water Alkalinity

Type: Site selection and site design

System: Estuarine, riverine, lacustrine, palustrine

Criterion: Use fine mineral sediments or soils containing high levels of aluminum or iron as the wetland substrate.

Rationale: Sediments containing high levels of aluminum, calcium, or iron favor the removal of phosphorous (Richardson, 1985) as does water with high alkalinity. Long term phosphorous absorption capacity of acidic wetlands is directly related to the extractable aluminum content of its soils.

Methods: For phosphorous removal, select a site which has primarily alluvial, alfisols, ferric, clay, or other fine soils. This is usually more critical in nontidal areas. For nitrogen removal, use highly organic soils. This is more critical in tidal areas.

To do this, first consult the U.S.D.A. soil survey for the soil series and phase on the site. If topsoil from an off-site location is to be used, determine the series and phase of this soil.

Subsequently, have a chemical analysis made of the topsoil for noncrystalline (amorphous) extractable aluminum content and percent organic carbon. These tests will determine the long-term phosphorous and nitrogen capacity of the soil.

Notes: At a minimum, in order for phosphates to be precipitated, it is essential that the water source for the wetland have an alkalinity higher than 20 mg/l. This test can easily be made by a water quality testing laboratory, or existing water quality information may be consulted.

Section 3.5
Channel Gradient and Water Velocity

Type: Site design

System: Estuarine, riverine, lacustrine, palustrine

Criterion: Create a gradual wetland basin gradient with low water velocity.

Rationale: Water velocity decreases with decreasing slope. As water velocity decreases, the potential for nutrient removal increases.

Methods: Keep the gradient of the wetland basin very gradual to enhance deposition and retention of sediment.

Use Table 2 to determine slope conditions which create depositional velocity conditions for channels less than 20 feet wide.

For channels greater than 20 feet, the peak annual flow velocity in the wetland basin should be less than 1.5 feet/second.

Table 2. Gradient necessary to create depositional velocity conditions.*

Mean Depth(ft)	Densely Wooded Floodplains	Densely Vegetated Emergent Wetlands not Totally Submerged by Floodflow	Moderately Vegetated or Totally Submerged Emergent Wetlands, or with Boulders	Unobstructed Channels
<0.5	<0.0250	<0.0100	<0.0038	<0.0018
0.5 - 1	<0.0150	<0.0060	<0.0023	<0.0012
1 - 2	-----------	<0.0030	<0.0012	<0.0006
2 - 3	-----------	<0.0017	<0.0006	<0.0003
3 - 4**	-----------	<0.0013	<0.0005	<0.0002
4 - 6**	-----------	<0.0008	<0.0003	<0.0001
6 - 8**	-----------	<0.0006	<0.0002	<0.0001
8 - 10**	-----------	<0.0004	<0.0002	----------
10 - 12**	-----------	<0.0003	<0.0001	----------

* Interpreted form SCS curves for channel flow and Manning's roughness coefficients.
** Assumes width, perpendicular to flow is less than 8 feet. If channel is 8 - 20 feet wide, the value in the row immediately below the value identified should be used.

Source: Adamus et al., 1987

Section 3.6
Outlet Characteristics

Type: Site design

System: Estuarine, riverine, lacustrine, palustrine

Criterion: Create a constricted surface water outlet, or no outlet at all.

Rationale: Wetlands with no outlets or with constricted outlets have an increased probability of sedimentation, adsorption, biological processing, and retention of nutrients.

Methods: *For wetlands with a gradient (channel flow):*
Generally, keep the width of the wetland outlet less than 1/3 the wetland's average width at ordinary high water, and/or keep the cross sectional area of the outlet less than 1/3 the cross sectional area of the inlet (see Figure 8).

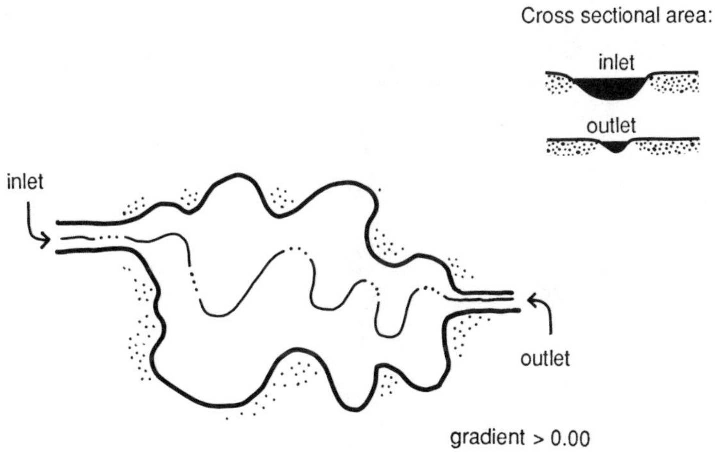

Figure 8. Wetland with channel flow.

For wetlands with no gradient or tidal wetlands:
Keep the total width of the outlet less than 1/10 the average width of the wetland; or have no outlet from the wetland (see Figure 9).

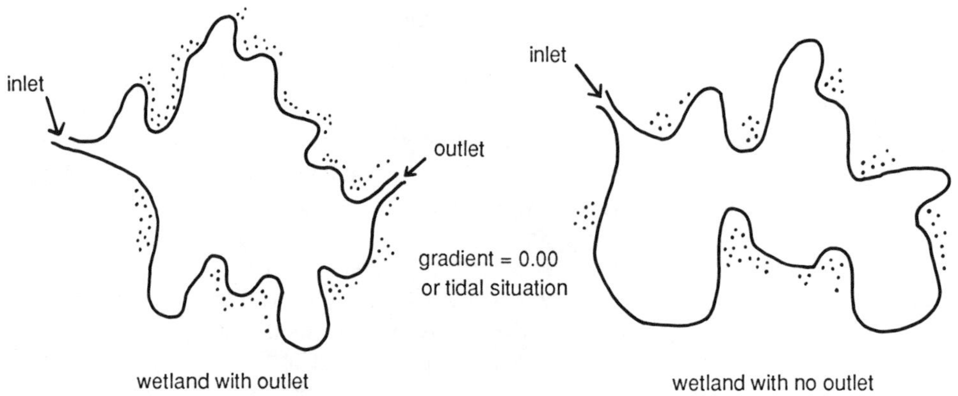

Detention times of several days or weeks desirable.

Figure 9. Wetland with no channel flow.

Section 3.7
Vegetation Class and Form Richness

Type: Site design

System: Estuarine, riverine, lacustrine, palustrine

Criterion: Include a vegetative cover which is predominantly forested, scrub/shrub, and/or persistent emergent. Emphasize vegetation diversity.

Rationale: All plants take up nutrients and store them in their tissues. The presence of vegetation also offers frictional resistance to water. This acts to bind the sediment and favors nutrient retention by burial.

Because different vegetation forms are involved in removing nutrients in different ways, planting a diversity of vegetation classes ensures that most nutrient cycling processes will be present.

Methods: In riverine locations, plant predominantly multi-stemmed woody vegetation. In other wetland systems, plant a combination of vegetation forms. Uptake is generally highest by emergent plants, particularly persistent species. These species may also transfer more nutrients below the zone of biological activity. Designs should emphasize this vegetation class, but also include a diversity of vegetation classes.

<u>Persistent emergents</u>
Bulrushes are particularly good for nitrogen removal (*Scirpus validus* or *S. lacustris*). Examples of other persistent emergent vegetation which are commercially available include the following:

Cattail	*Typha spp.*
Iris	*Iris pseudacorus* or *I. versicolor*
Rush	*Juncus spp.*
Cordgrass	*Spartina spp.*

Reedgrass	*Calamagrostis spp.*
Sawgrass	*Cladium jamaicense*
Rice cutgrass	*Leersia oryzoides*
Switch-grass	*Panicum virgatum.*

Other persistent emergent species may be applicable, depending on the geographic location of the replacement wetland.

For woody plants, specify multi-stemmed as opposed to single stemmed when ordering from the nursery supplier. A partial list of woody plants includes the following (U.S.D.A. Soil Conservation Service, 1977):

Multi-stemmed Woody Plants

Alder	*Alnus spp.*
Buttonbush	*Cephalanthus occidentalis*
Viburnum	*Viburnum spp.*
Crabapple	*Malus spp.*
Chokecherry	*Prunus virginiana*
Purple osier willow	*Salix purpurea*
Winterberry	*Ilex verticillata*
Inkberry	*Ilex glabra*
Water willow	*Decodon verticillatus*
Gray stem dogwood	*Cornus racemosa*
Silky dogwood	*Cornus amomum*
Red osier dogwood	*Cornus stolonifera*
Northern bayberry	*Myrica pensylvanica*
Streamco willow	*Salix purpurea - "streamco".*

Other woody species may be applicable, depending on geographic location. In most cases, native nursery grown stock or seed is desirable. Contact the local SCS office and State highway department for a listing of erosion control plants which tolerate wet conditions. Also, contact local nurseries which specialize in native wetland plants. A partial listing of wetland and native plant suppliers may be found in Appendix B.

Notes: Aquatic bed vegetation also removes nutrients seasonally but does not offer significant frictional resistance to suspended sediments.

Section 3.8
Vegetated Width

Type: Site design

System: Estuarine, riverine, lacustrine, palustrine

Criterion: Use wide stands of multi-stemmed woody and/or persistent emergent vegetation.

Rationale: Extensive stands of vegetation in a wetland act to slow water velocity. When velocity is slowed, nutrients in the water are more likely to sink to the soil substrate.

Methods: *For wetlands without a constricted outlet:*
Plant wide stands of vegetation. To maximize this function, vegetation should completely span the wetland or have an average total width greater than 500 feet. These plants may include multi-stemmed woody and persistent emergent plants where no visible surface water will be present or emergent plants only where surface water will be present. Water depth should never exceed 50 percent of plant height.

For wetlands with a constricted outlet:
Where visible standing water will be present, plant a high density of persistent emergent plants (greater than 50 stems/square meter) throughout the wetland. Water depth should never exceed 50 percent of plant height.

Section 3.9
Wetland/Watershed Ratio

Type: Site selection

System: Estuarine, riverine, lacustrine, palustrine

Criterion: Keep a low wetland to watershed ratio.

Rationale: The larger the watershed, the greater the amount of suspended sediment and nutrients likely to enter a wetland. Nutrient delivery generally increases with increasing watershed size.

Methods: Select a replacement site which comprises less than five percent of the size of the watershed. Additionally, select a watershed where few other wetlands are located upslope of the wetland replacement site.

Chapter 4
Sediment/Toxicant Retention

Description of the Function:

Sediments frequently contain chemically and physically attached nutrients and contaminant materials, such as heavy metals, pesticides and other organic toxicants. Sediments and associated toxicants are carried by runoff or channel flow into wetlands, where they can be removed temporarily or permanently from the water column by sediment deposition. Nutrients and toxicants carried by sediments into the wetland can be removed by burial, chemical breakdown, and/or assimilation into plant and animal tissues. Sediments may also be temporarily retained by a wetland before moving further downstream.

General Design Concepts:

The principal factor affecting a wetland's ability to trap sediments is the change in the velocity or energy level of incoming water. Decreased water velocity results in sediment deposition.

A wetland which retains water for long periods of time or indefinitely is ideal for sediment trapping. The ability of a wetland to retain water is determined by the following physical and biological factors:

- A wetland with a constricted outlet or no outlet will slow water and hold it in the basin (Section 4.3).

- A gentle gradient in the wetland basin will slow water velocity (Section 4.2).

- Dense wetland vegetation will act to slow water velocity, to force water to flow through a longer course, to retain it longer in the basin, and to discourage resuspension of bottom sediments (Sections 4.8 and 4.9).

- A long duration and extent of seasonal flooding allows for a longer water retention time (Section 4.7).

- Shallow water depth increases frictional resistance and slows water velocity. Vegetation which persists throughout the year is optimal for this function (Section 4.5).

Once the sediments are settled, toxicants may be bound to the substrate. This process is especially true of wetlands with underlying organic soils, which may permanently complex with metals and synthetic organic toxicants (Section 4.13).

Minimal fetch and exposure of the wetland to wind and wave action will discourage the resuspension and transport of sediment out of the wetland, acting to retain sediments for long periods of time or indefinitely (Section 4.4).

The wetland system is also important in determining the sediment trapping ability. Estuarine wetlands which support a mixohaline salinity encourage the flocculation of clays in sediment at the fresh water/salt water interface. In contrast, riverine systems are more likely to carry large quantities of suspended sediments and associated toxicants and therefore not function well as sediment traps. Palustrine and lacustrine wetlands, because they are more likely to be basin-shaped, offer good sediment trapping capability (Section 4.1).

Finally, it is important for the water source of the wetland to be generated principally from surface runoff, the principal source of sediment laden water (Section 4.6).

The remaining features relate to the selection of the wetland replacement site itself, as the inflowing waters must contain sediments in order for a wetland to trap them. These features include the selection of a watershed which contains sediment producing sources and substrate type (Sections 4.11, 4.12, and 4.13).

Specific Site Selection and Site Design Features:

The following table describes the features used in the Sediment/Toxicant Retention function. With the exception of Section 4.6, Water Source, these features were derived from the WET predictors.

Generally, estuarine, lacustrine, and palustrine systems are recommended for this function. Riverine systems do not retain sediment long enough to perform this function satisfactorily.

However, sediment and toxicant loading rates should preferably not exceed the wetland's ability to assimilate these substances without causing ecological harm or threatening or conflicting with other wetland functions (see Table 12).

Table 3. Sediment/toxicant retention site selection and site design features.

Section	Feature	System [1]	Site Selection	Site Design	Importance to Function [2]	Notes [3]
4.1	Wetland System	E,L,P	X		Moderate	10
4.2	Channel Gradient and Water Velocity	E,L,P		X	High	7, 41 renamed
4.3	Outlet Characteristics	E,L,P		X	High	8,9 renamed
4.4	Fetch/Exposure	E,L,P	X		Moderate	19
4.5	Water Depth	E,L,P		X	Moderate	43
4.6	Water Source	E,L,P	X	X	High	new
4.7	Flooding Extent and Duration	L,P	X	X	Moderate	35
4.8	Vegetated Width/Vegetation Class	E,L,P		X	High	12, 36 renamed
4.9	Water/Vegetation Proportions and Interspersion	E,L,P		X	Moderate	15.1, 31 renamed
4.10	Wetland/ Watershed Ratio	E,L,P	X		High	5
4.11	Land Cover of the Watershed	E,L,P	X		Moderate	21
4.12	Sediment and Contaminant Sources	E,L,P	X		Moderate	25, 27 renamed
4.13	Substrate Type	E,L,P	X	X	High	45

1 E - estuarine, L - lacustrine, P - palustrine

2 These ratings are generally derived from Volume I of WET (FHWA-IP-88-029). Some values were modified in relation to wetlands replacement criteria by Paul Adamus (U.S. EPA Environmental Research Lab, Corvallis, Oregon).

3 This column describes the WET 2.0 predictors used in deriving each feature. If only a number is given, it refers to the predictor number in WET 2.0. RENAMED means that the title of the WET 2.0 predictor was modified to show the new feature. NEW means the feature is not based on any given WET 2.0 predictor, but was inferred from other information.

WET 2.0 Predictors Not Used

1 Climate; 13 Vegetation Class/Subclass; 22 Flow, Gradient, Deposition; 28 Direct Alteration; 34 Water Level Control; 42 Velocity (Secondary); 48 Salinity and Conductivity; 49 Aquatic Habitat Features (Riverine); 55 Suspended Solids; 64 Total Suspended Solids

Section 4.1
Wetland System

Type: Site selection

System: Estuarine, lacustrine, palustrine

Criterion: Avoid riverine wetland systems.

Rationale: Sediment retention times are generally least in riverine wetlands. Flooding events frequently erode bottom sediments. Riverine systems also may carry large quantities of suspended sediments and associated toxicants.

Conversely, mixohaline waters (0.5 to 18 ppt salinity) and waters with high conductivity (greater than 500 micromhos/cm) encourage flocculation and settling of clay particles. Estuarine systems which support salinities in this range are most desirable for sediment and toxicant retention. Lacustrine and palustrine wetlands are also generally favorable.

Methods: Select an estuarine, lacustrine, or palustrine system.

Section 4.2
Channel Gradient and Water Velocity

Type:	Site design
System:	Estuarine, lacustrine, palustrine
Criterion:	Keep the gradient of the wetland gradual, with low water velocity.
Rationale:	Water velocity decreases with decreasing slope. As water velocity decreases, the potential for sediment and toxicant retention and deposition increases.
Methods:	Keep the gradient of the wetland very gradual to enhance deposition and retention of sediment. For methods, see Section 3.5, Nutrient Removal/Transformation.

Section 4.3
Outlet Characteristics

Type: Site design

System: Estuarine, lacustrine, palustrine

Criterion: Use a constricted surface water outlet, or no outlet at all.

Rationale: Wetlands which restrict the outflow of water are more likely to retain absorbed sediments and toxicants. The longer the retention time of the wetland, the more the sediments are likely to settle. Lighter particles, such as colloidal clays, have the longest sedimentation rate. Inorganic particles are heaviest and therefore settle out fastest.

Methods: Sediment retention in wetlands without outlets is usually complete, as the flow of water is stopped, and all particles regardless of size can settle out. Wetlands without outlets are therefore more favorable than those with constricted outlets.

For methods, see Section 3.6, Nutrient Removal/Transformation.

Notes: In developing the design for this type of wetland, it will be critical to determine the potential for flood damage to adjacent areas.

Section 4.4 Fetch/Exposure

Type:	Site selection
System:	Estuarine, lacustrine, palustrine
Criterion:	Locate the wetland in a sheltered area where the adjacent topographic relief is sufficient to protect the site from wind.
Rationale:	Sheltered areas are less prone to wind mixing, which encourages the suspension and transport of sediments out of the wetland.
Methods:	Locate the wetland so that adjacent topographic relief or adjacent vegetation is sufficient to shelter the wetland from wind. Select a location where the open water fetch adjoining the wetland is less than two miles (see Figure 10). A windbreak size may be determined by the height of the vegetation and/or relief multiplied by the length of the vegetation and/or relief parallel to the wetland. A minimum size is 2,000 square feet.

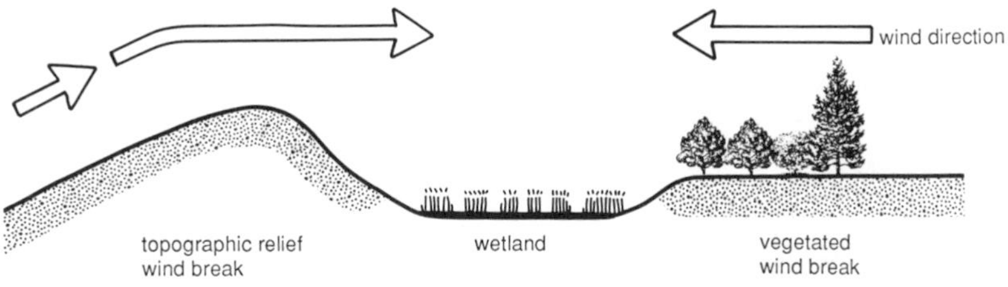

Adjacent topographic relief and/or vegetation may help shelter the wetland from wind mixing of sediments.

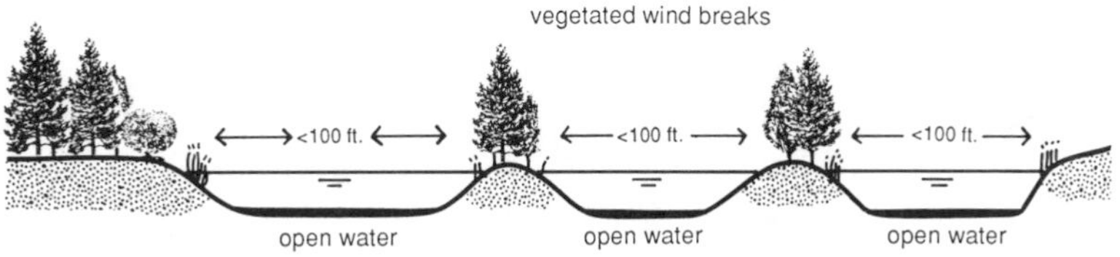

A wetland whose open water expanse is broken by vegetated and/or topographic wind breaks will be less likely to have sediment resuspension problems. Coniferous trees serve this purpose, as they will provide a wind screen all year, as opposed to seasonally.

Figure 10. Topographic and vegetative treatments to minimize fetch and exposure effects.

Section 4.5
Water Depth

Type:	Site design
System:	Estuarine, lacustrine, palustrine
Criterion:	Provide predominantly shallow water depths.
Rationale:	Wetlands with shallow water offer greater frictional resistance to flow which in turn affects suspended solids. Frictional resistance (velocity reduction) favors sedimentation.
Methods:	The predominant water depth of the wetland should be less than 40 inches except where the threat of wind resuspension of sediments is great (see Section 4.4, Fetch/Exposure). Shallow water also favors vegetation, as described in Section 4.8, Vegetated Width/Vegetation Class.

Section 4.6
Water Source

Type: Site selection and site design

System: Estuarine, riverine, lacustrine, palustrine

Criterion: Use a surface water source as the wetland's principal water supply.

Rationale: Water supply originating from a surface water source, such as channel flow, overland flow, and precipitation, will carry sediments into the wetland from the watershed. In comparison, ground water sources are less likely to contain high levels of nutrients and thus provide less chance to perform this function. The wetland may therefore have any type of inlet (none, intermittent, or permanent) as long as it is receiving surface water inflow.

Methods: Generally, avoid using excavated basins which intersect the water table as a principal water source for the wetland.

See Section 3.2, Nutrient Removal/Transformation for methods.

Section 4.7
Flooding Extent and Duration

Type: Site selection and site design

System: Lacustrine, palustrine

Criterion: Select a site adjacent to surface water which is subject to seasonal flooding, or a site which experiences a seasonal high water table.

Or, create a wetland whose surface waters expand significantly in acreage on a seasonal basis.

Rationale: The greater the seasonal extent of flooding, the greater the potential for settlement of suspended sediments and associated toxicants.

Methods: For site selection, locate a site which experiences a significant rise in flood level on a seasonal basis.

For site design, consult Chapter 2, Wetland Hydrology, for wetlands using a surface water supply. The cross-sectional area of the stream and its associated wetland must accommodate flood waters for the specified extent and duration identified in Figure 11.

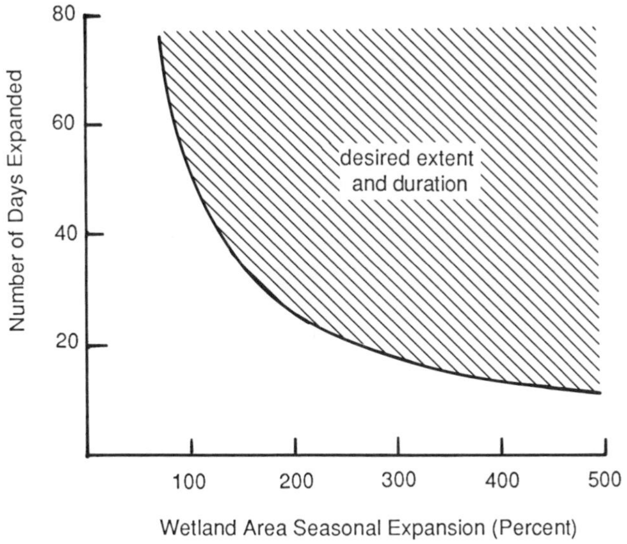

Figure 11. Criteria for significantly expanded seasonally flooded surface area.
(Source: modified from Adamus et al., 1987).

Section 4.8
Vegetated Width/Vegetation Class

Type: Site design

System: Estuarine, lacustrine, palustrine

Criterion: Plant wide stands of multi-stemmed woody and/or persistent emergent vegetation.

Rationale: Extensive stands of vegetation offer frictional resistance to water flow, enhancing sedimentation. The wider the stand of vegetation, the greater the potential to encourage sedimentation.

In addition, wetland vegetation reduces the resuspension of bottom sediments from wind mixing and lengthens the flow path of water through the wetland.

Wetland vegetation also contributes to the organic content of the bottom sediments which in turn helps retain toxicants associated with sediments.

Methods: See methods discussed in Section 3.8, Nutrient Removal/Transformation.

Section 4.9
Water/Vegetation Proportions and Interspersion

Type: Site design

System: Estuarine, lacustrine, palustrine

Criterion: Maximize the percent of emergent vegetation with respect to the percent of open water. Also, create low vegetation/water interspersion conditions.

Rationale: Vegetation offers frictional resistance to flowing water, encouraging settlement of sediments and discouraging resuspension. Dense vegetation also decreases the probability of sediment resuspension by wind or wave action.

Methods: In the standing and flowing water portions of the wetland, maximize the density of persistent emergent vegetation and reduce the area of open water pools and channels.

Because persistent species remain standing throughout the winter season, they will function to remove sediment throughout the year. The design should therefore emphasize the use of persistent emergent species.

Section 4.10
Wetland/Watershed Ratio

Type:	Site selection
System:	Estuarine, lacustrine, palustrine
Criterion:	Keep a high wetland to watershed ratio.
Rationale:	The larger the wetland relative to the watershed, the greater the proportional amount of suspended sediment likely to be retained. Sediment delivery to the wetland generally increases with increasing watershed size and fewer alternative storage areas upslope of the replacement site.
Methods:	Select a replacement site which comprises more than five percent of the size of the watershed. Additionally, select a watershed where few other wetlands are located upslope of the wetland replacement site.

Section 4.11
Land Cover of the Watershed

Type: Site selection

System: Estuarine, lacustrine, palustrine

Criterion: Select a site whose watershed is predominantly urban, agricultural, and/or disturbed land.

Rationale: Runoff from exposed urban and agricultural soils is more likely to carry sediment, and toxicants associated with them, providing an opportunity for wetlands to protect receiving waters farther downstream.

Methods: Locate the replacement site in a watershed comprised primarily of agricultural and/or urban land. Obtain a recent aerial photograph of the prospective replacement site and its watershed to determine this factor.

Section 4.12
Sediment and Contaminant Sources

Type: Site selection

System: Estuarine, lacustrine, palustrine

Criterion: Select a watershed which will contribute inorganic sediment and/or waterborn contaminants to the wetland.

Rationale: Wetlands that receive runoff from watersheds with erosion prone areas or contaminant sources are most likely to have an opportunity to function as toxicant and sediment removers. Erodible soils and steep slopes are conducive to providing and transporting soil solids into an adjacent wetland for sediment removal.

Methods: Select a replacement site which would receive runoff from one or more of the following sources:

- stormwater outfall,
- surface mines,
- exposed soils,
- irrigation return waters,
- soil series categorized by the U.S.D.A. Soil Conservation Service as constituting an erosion hazard,
- sand and gravel pits,
- gullies and severely eroding streambanks,
- industrial or sewage outfall,
- landfills,
- pesticide treated areas,
- heavily traveled highways, and
- urban runoff.

Other site selection criteria include the following:

- unstable slopes steeper than 10 percent (steeper than one percent if alluvial clay) immediately adjacent to the site, and
- channelized tributaries immediately upstream of the replacement site.

Section 4.13
Substrate Type

Type: Site selection and site design

System: Estuarine, lacustrine, palustrine

Criterion: Select a site which contains predominantly organic soil or use predominantly organic soil from an off-site location as a substrate.

Rationale: Toxicant retention, especially of metals and synthetic organic materials (such as PCBs) is associated with organic soils.

Methods: Select a site which contains peat or muck (organic) soils as a wetland substrate. If these are not available, fine mineral soils high in organic content are then recommended. The least attractive substrate types are sand, cobble-gravel, rubble, and bedrock.

If an organic substrate is not available on site, locate an off-site source. Apply organic soil as a topsoil 6 to 12 inches thick to the replacement site.

Soils from the wetland area designated to be filled may be stockpiled and used for the replacement wetland's substrate. These soils are often high in organic content or contain fine mineral soils. However, a soil analysis should be made to determine these characteristics prior to use.

Chapter 5
Shoreline Stabilization

Description of the Function:

Shoreline stabilization is the binding of soil at the shoreline or water's edge by wetland plants, and the physical dissipation of erosive energy caused by waves, currents, tides, or ice in a basin or channel. Shoreline stabilization by wetlands protects adjacent uplands from erosion, thereby protecting adjacent land uses. Stabilization also has the net effect of minimizing the deposition of the eroded sediment in navigable channels.

General Design Concepts:

The frictional resistance a wetland offers to erosive energy depends on the vegetated width of the wetland (Section 5.4), the density of vegetation (Section 5.5), and the height of the vegetation relative to incoming waves and currents (Section 5.3). Persistent emergent and woody vegetation are preferred because they offer frictional resistance throughout the year (Section 5.3).

Frictional resistance to high energy water may also be created by a wetland whose morphology is flat or nearly flat (Section 5.2). This forces water to spread out over a large area rather than concentrating the erosive forces in a small area.

For a wetland to be valued as a shoreline stabilizer, potentially erosive conditions must be present (Section 5.1). These may take the form of flowing water, a long fetch adjacent to eroding areas, and water with low turbidity (Section 5.6). Although these conditions enhance a wetland's importance, their potential for threatening the self-sustainability of newly planted wetlands must be considered.

Specific Site Selection and Site Design Features:

Table 4 describes the site selection and site design features for Shoreline Stabilization. All features were derived from WET predictors, with the exception of Section 5.7, Shoreline Geometry, and Section 5.8, Resource Protection. The features may be applied to all wetland systems.

Table 4. Shoreline stabilization site selection and site design features.

Section	Feature	System [1]	Site Selection	Site Design	Importance to Function [2]	Notes [3]
5.1	Erosive Conditions	E,R,L,P	X		Moderate	25, 34, 41.2 renamed
5.2	Sheet Flow	E,L,P		X	Moderate	15.2 renamed
5.3	Vegetation Class	E,R,L,P		X	Moderate	12
5.4	Vegetated Width	E,R,L,P		X	High	36
5.5	Water/Vegetation Proportions	E,R,L,P		X	High	31
5.6	Fetch/Exposure	E,R,L,P	X		High	19
5.7	Shoreline Geometry	E,R	X	X	High	new
5.8	Resource Protection	E,R,L,P	X		Moderate	new

1 E - Estuarine, R - Riverine, L - Lacustrine, P - Palustrine

2 These ratings are generally derived from Volume I of WET (FHWA-IP-88-029). Some values were modified in relation to wetlands replacement criteria by Paul Adamus (U.S. EPA Environmental Research Lab, Corvallis, Oregon).

3 This column describes the WET 2.0 predictors used in deriving each feature. If only a number is given, it refers to the predictor number in WET 2.0. <u>RENAMED</u> means that the title of the WET 2.0 predictor was modified to show the new feature. <u>NEW</u> means the feature is not based on any given WET 2.0 predictor, but was inferred from other information.

<u>WET 2.0 Predictors Not Used</u>

7 Gradient; 15.1 Vegetation Interspersion; 22 Flow, Gradient, Deposition; 41 Velocity (Spatially Dominant); 45 Substrate Type

Section 5.1
Erosive Conditions

Type: Site selection

System: Estuarine, riverine, lacustrine, palustrine

Criterion: Locate the wetland where the protection of streambanks, shores, or adjoining properties is desired.

Rationale: In order for a wetland to offer shoreline stabilization value to society, it must be exposed to erosive forces.

Methods: If the wetland's permanency can be assured, locate the wetland where one or more of the following conditions are present in the adjacent area:

- flowing water with velocities exceeding 1.5 feet/second,
- boat wakes,
- open water expanse greater than 100 feet across, but less than 1.2 miles (see Section 5.7), and
- unstable slopes exceeding 10 percent immediately adjacent to the wetland site, channelized tributaries immediately upstream, or large impoundments exceeding 20 feet in height at the outlet.

Existing eroded banks and shorelines indicate that one or more of the above erosive forces are present.

Section 5.2
Sheet Flow

Type: Site design

System: Estuarine, lacustrine, palustrine

Criterion: Water should flow through the wetland as sheet (unchannelized) flow.

Rationale: Frictional resistance is higher when water spreads out over a large area, rather than being confined to a channel. When frictional resistance is high, potential erosiveness is lower.

Methods: Design a wetland whose morphology allows water to spread out rather than remain in a channel. A topographically flat bottom will lend itself to sheet flow conditions.

Velocity conditions less than 0.3 feet/second are recommended for wetlands greater than 20 feet wide.

Section 5.3
Vegetation Class

Type: Site design

System: Estuarine, riverine, lacustrine, palustrine

Criterion: Plant one or more of the following wetland vegetation classes: forested, scrub/shrub, and persistent emergent.

Rationale: Plants that dissipate erosive forces by creating frictional drag are rigid, persistent, and tall enough to penetrate the entire water column during seasonal flooding. In addition, plants anchor shorelines and floodplain sediments by binding soil with their roots (Seibert, 1968).

Methods: Select a hydrologic regime which will specifically support forested, scrub/shrub, and/or emergent vegetation. For emergent areas, select persistent species only.

A partial listing of wetland and native plant suppliers may be found in Appendix B.

Section 5.4
Vegetated Width

Type: Site design

System: Estuarine, riverine, lacustrine, palustrine

Criterion: Use wide stands of wetland vegetation near the shoreline.

Rationale: Vegetation offers frictional resistance to flowing water, and hence, acts to dissipate its erosive forces.

Methods: Plant wide stands of vegetation consisting of persistent emergent, scrub/shrub, or forested vegetation. To maximize for this function, the average width of vegetation should be greater than 30 feet total, perpendicular to flow. Average water depth should not exceed 50 percent of plant height at the time of establishment (Knutson et al., 1981).

Section 5.5
Water/Vegetation Proportions

Type: Site design

System: Estuarine, riverine, lacustrine, palustrine

Criterion: Avoid unvegetated, open water. In areas supporting surface water, ensure a high density of persistent emergent vegetation.

Rationale: Persistent emergent vegetation will provide shoreline stabilization by offering frictional resistance to waves and by binding the soil with its roots.

Methods: If portions of the wetland support a semipermanently flooded, permanently flooded, or intermittently exposed hydroperiod (zones where surface water occurs), plant dense stands of persistent emergent vegetation so there are few unvegetated open water areas.

Avoid unvegetated, open water. This includes water that is too deep to support persistent emergent vegetation (generally depths greater than two feet).

Notes: Generally, trees and shrubs bind soil better than persistent emergent vegetation. These two vegetation classes should therefore dominate the wetland (see Section 5.3, Vegetation Class). Persistent emergent vegetation should be included where semi-permanent, permanent, and intermittently exposed conditions are present.

Section 5.6
Fetch/Exposure

Type: Site selection

System: Estuarine, riverine, lacustrine, palustrine

Criterion: Locate the site in an area which is perpendicular to the dominant wind direction and has an open water fetch greater than 100 feet, but less than 1.2 miles.

Rationale: Erosive forces, such as waves and storm surges, are more likely to be amplified if fetch is great. Wetlands that intercept waves and protect adjacent shorelines are therefore more likely to provide needed shoreline stabilization if they are first well established.

However, areas with extreme fetch are poor candidates for the establishment of intertidal vegetation. Areas exposed to high waves on a frequent basis are too physically unstable.

Methods: Locate the wetland where open water fetch is greater than 100 feet, and in a location which is perpendicular to the dominant wind direction.

Both average fetch and longest fetch should be measured at the replacement site (Figure 12).

Average Fetch. Measure the distance of open water perpendicular to replacement site and 45° either side of the perpendicular. The average of these three measurements should not exceed 0.6 miles.

Longest Fetch. Measure the distance of open water perpendicular to the replacement site and 45° either side of the perpendicular. The longest of these three measurements should not exceed 1.2 miles.

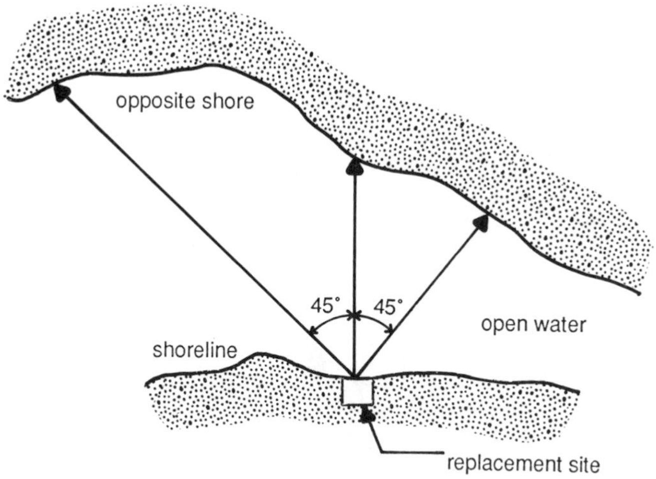

Figure 12. Measurement of fetch (Source: Knutson et al., 1981).

Section 5.7
Shoreline Geometry

Type: Site selection and site design

System: Estuarine, tidal riverine

Criterion: Locate or design the replacement site so that it's geometry will effectively shelter it from waves.

Rationale: Sites located in narrow coves may be effectively sheltered from waves approaching at oblique angles and will be subjected to large waves only when winds blow directly onshore or nearly so. Conversely, sites located on headlands are exposed to waves from many directions.

Methods: Determine the general shape of the potential replacement site in relation to its immediate shoreline. To do this, survey the replacement site's shoreline and up to 700 feet of shoreline to either side.

Several basic shore configurations are possible: coves, meandering shorelines, and headlands (see Figure 13). The cove configuration is most desirable and the headland configuration least desirable, with the meandering shoreline being intermediate.

The shoreline's existing configuration must contain a cove configuration, or one may be created in the design process by physically modifying the shoreline.

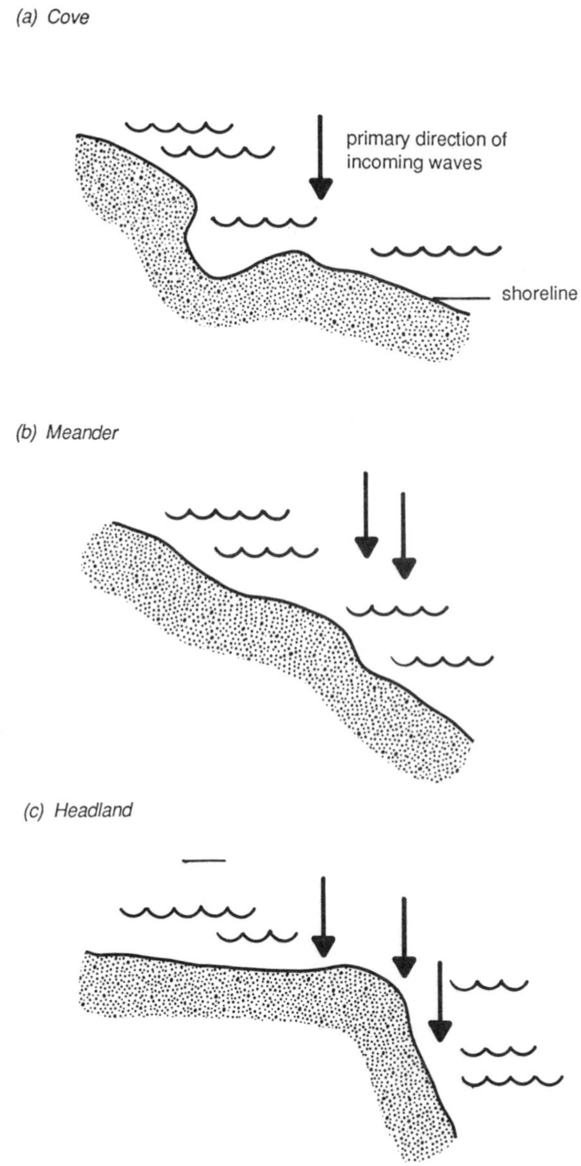

Figure 13. Types of shoreline geometry.

Section 5.8
Resource Protection

Type: Site selection

System: Estuarine, riverine, lacustrine, palustrine

Criterion: Locate the wetland adjacent to a resource requiring protection from sedimentation.

Rationale: The stabilization of shore environments whose adjacent upland supports one or more significant natural or social values is particularly important. Also, many wetlands stabilize shorelines by aiding in the accretion of soil which would otherwise be deposited in nearby navigable channels.

Methods: Determine if one or more of the following are located adjacent to the replacement area:

- significant wildlife areas,
- fish hatcheries, nurseries, shellfish beds, and
- dwellings or other buildings.

Chapter 6
Floodflow Alteration

Description of the Function:

Floodflow alteration is the process by which peak flows from runoff, surface flow, and precipitation are stored or delayed. Wetlands, as well as upland areas, act to detain flood waters by intercepting sheet flow and flood waters. By lowering flood peaks, wetlands act to decrease flood-related damage.

The importance of a wetland in altering floodflows depends to a great extent on its position in the watershed and its outlet characteristics. Wetlands located in the upper portion of the watershed are most effective if the total acreage of wetlands and other surface waters above them in the watershed is less than about seven percent of the watershed(Ogawa and Male, 1983). Wetlands low in the watershed can be effective regardless of the available upstream storage. However, wetlands can also aggravate flooding by impeding flow. Watershed-specific hydrologic modeling is recommended to determine if a wetland constructed in a new location could pose a liability.

The storage capacity of a wetland depends primarily on the type and location of outlets. Wetlands with no outlet will store all incoming water. If a constricted outlet is present, significant water storage or detention will also occur. Even where no constriction exists (e.g. floodplains), wetlands may desynchronize flow by soil capillary storage and frictional roughness of vegetation.

General Design Concepts:

In selecting the site for wetland design, the location and size of other wetlands in the watershed and the potential for runoff to reach the wetland should be considered (Section 6.5). Watersheds with a relatively small portion of their upslope area occupied by wetlands will benefit most in terms of effective floodflow alteration. A watershed with greater than 10 percent of its acreage in wetlands may not realize an incrementally significant floodflow alteration benefit from additional wetland acreage (Ammon et al., 1981).

In addition, a watershed which produces high volumes of runoff will benefit most from the addition of wetland area. Impervious surfaces, including soils with slow infiltration rates, steep slopes, and developed areas covered by asphalt or concrete, will generate more runoff than undeveloped land with vegetative cover (Sections 6.3 and 6.4).

Flood water storage is most significant in palustrine, lacustrine, and upper riverine wetland systems. Other wetland systems, including estuarine and lower riverine, will be more affected by tidal influence and may actually act to aggravate floodflows (Section 6.1).

The storage capacity of the wetland is also a function of the wetland morphology. If appropriate, the wetland may be located high enough in the watershed and be of sufficient proportional size to store all stormwater (Section 6.5). This wetland would have no outlet and would retain water until it is lost to the ground water table or lost by evapotranspiration. In situations where this is not practical, a constricted outlet may be incorporated into the design (Section 6.2). A constricted outlet as well as frictional resistance in the channel and the wetland will slow the water flow and desynchronize the peak flow.

Desynchronization of headwater runoff will increase the duration of the flood event but will reduce the effects of the peak flow on downstream areas. Frictional resistance in the wetland may be created by dense vegetation or other obstructions to flow, including boulders and logs (Sections 6.6, 6.7, and 6.8).

Specific Site Selection and Site Design Features:

It is recommended that the same design process be used as for a conventional stormwater detention or retention basin design. This procedure includes a determination of incoming flow volumes, storage volumes, and outlet sizes. A hydrologic analysis will also determine the size and morphology of the wetland basin which is critical in attenuating floodflow. Subsequently, many of the features described in this chapter may be incorporated into a conventional stormwater detention basin.

Methods for flood routing and sizing detention/retention basins may be found in the following publications:

- *National Engineering Handbook,-Hydrology,* Section 4. (U.S.D.A. Soil Conservation Service. 1985).

- *Urban Hydrology for Small Watersheds.* (U.S.D.A. Soil Conservation Service, 1975).

Many other publications are available, some being specific to a particular region of the country.

Table 5 describes the parameters used in the floodflow alteration function. All of the features were derived from the WET predictors. Several features combine two or more predictors into one site design or selection feature. These features apply to all the wetland systems except estuarine.

Table 5. Floodflow alteration site selection and site design features.

Section	Feature	System [1]	Site Selection	Site Design	Importance to Function [2]	Notes [3]
6.1	Wetland System	R,L,P	X		Moderate	10
6.2	Outlet Characteristics	R,L,P		X	High	8, 9 renamed
6.3	Land Cover of the Watershed	R,L,P	X		Moderate	21
6.4	Watershed Soils	R,L,P	X		Moderate	24.4
6.5	Wetland/ Watershed Ratio	R,L,P	X		High	5
6.6	Water/Vegetation Proportions and Interspersion	R,L,P		X	Moderate	15.1, 31 renamed
6.7	Vegetation Class	R,L,P		X	Moderate	12
6.8	Sheet Flow	R,L,P		X	Moderate	15.2

1 R - riverine, L - lacustrine, P - palustrine

2 These ratings are generally derived from Volume I of WET (FHWA-IP-88-029). Some values were modified in relation to wetlands replacement criteria by Paul Adamus (U.S. EPA Environmental Research Lab, Corvallis, Oregon).

3 This column describes the WET 2.0 predictors used in deriving each feature. If only a number is given, it refers to the predictor number in WET 2.0. RENAMED means that the title of the WET 2.0 predictor was modified to show the new feature. NEW means the feature is not based on any given WET 2.0 predictor, but was inferred from other information.

WET 2.0 Predictors Not Used
1 Climate; 2 Acreage; 11 Fringe or Island; 22 Flow, Gradient, Deposition; 23 Ditches/Canals/Channelization/ Levees; 32 Hydroperiod; 35 Flood Extent and Duration; 63 Discharge Differential

Section 6.1
Wetland System

Type: Site selection

System: Nontidal riverine, lacustrine, palustrine

Criterion: Select a nontidal wetland system.

Rationale: The effect of tidal wetlands on floodflow alteration is usually inconsequential because of the fluctuating water level conditions. However, tidal wetlands in the upper reaches of the tidal influence zone can be important flood storage areas.

Nontidal riverine, lacustrine, and palustrine wetlands are more likely to provide the needed storage capacity to offset flood peaks.

Methods: Select a site which will not be hydrologically influenced by tides. Flood storage is most significant in palustrine, lacustrine, and upper riverine locations.

Section 6.2
Outlet Characteristics

Type: Site design

System: Nontidal riverine, lacustrine, palustrine

Criterion: Do not include a permanent outlet in the wetland. If an outlet is present, it should be a constricted outlet.

Rationale: Wetlands with no outlet or a constricted outlet will store water for a longer period of time than wetlands with outlets. Wetlands without outlets will store all incoming water. A constricted outlet will also allow for the storage of water which would otherwise be lost to downstream surface water.

Methods: A wetland designed for flood storage should be based on a particular flood or storm frequency (2 year, 5 year, etc.). Size the storage volume of the wetland to handle the conditions of the selected flood or storm frequency. The wetland should be of sufficient size relative to the inflow so that outflow or overflow of the basin occurs only on an intermittent basis.

If the surface water inflow is too great to be contained within the wetland, a constricted permanent outlet can be designed. For wetlands with channel flow, a constricted outlet should be less than 1/3 the average width of the wetland at the time of the selected flood or storm event or the cross sectional area of the outlet should be less than 1/3 the cross sectional area of the inlet.

For wetlands with no channel flow, the width of the outlet should be less than 1/10 the average width of the wetland.

See Section 3.6, Nutrient Removal/Transformation.

Section 6.3
Land Cover of the Watershed

Type: Site selection

System: Nontidal riverine, lacustrine, palustrine

Criterion: Select a watershed which has relatively large areas of impervious surfaces.

Rationale: Greater volumes of runoff and higher flood peaks are produced in watersheds having primarily impervious surfaces. A wetland located in a watershed with high volumes of runoff from impervious surfaces would have greater potential benefit to society if it attenuates floodflows.

Methods: Aerial photography of the watershed may be used to determine the extent of impervious surfaces. Select a watershed which consists principally of land covered in asphalt or concrete and where lot sizes are less than 1/4 acre.

Section 6.4
Watershed Soils

Type: Site selection

System: Nontidal riverine, lacustrine, palustrine

Criterion: Select a watershed which has relatively large areas of impermeable soils.

Rationale: Greater runoff and higher flood peaks are produced in watersheds having primarily impermeable soils. These types of soils impede infiltration of water and therefore produce increased runoff. Wetlands located downslope in watersheds supporting these conditions are more likely to provide flood attenuation.

Methods: Using a soil survey, determine the infiltration characteristics of soils in the watershed where the replacement site is located. Soils can be evaluated using the local soil survey, which describes the infiltration rate in inches per hour for each soil series.

Four infiltration rates have been classified by the National Cooperative Soil Survey:

- *Very low.* Soils with infiltration rates of less than 0.25 cm (0.1 in.) per hour; soils in this group are very high in percentage of clay.
- *Low.* Infiltration rates of 0.25-1.25 cm (0.1-0.5 in.) per hour; most of these soils are shallow, high in clay, or low in organic matter.
- *Medium.* Infiltration rates of 1.25-2.5 cm (0.5-1.0 in.) per hour; soils in this group are loams and silts.
- *High.* Rates of greater than 2.5 cm (1.0 in.) per hour; these are deep sands, deep well-aggregated silt loams, and some tropical soils with high porosity.

Section 6.5
Wetland/Watershed Ratio

Type: Site selection

System: Nontidal riverine, lacustrine, palustrine

Criterion: Select a location within a watershed which presently supports few other wetlands.

Rationale: The higher the proportion of wetlands upslope in the watershed, the less effect an additional wetland will have on floodflow alteration. Studies indicate that 50 percent of flood peak reduction results from the first 5 percent of wetland area in the watershed (Novitzki, 1979). The incremental gain in terms of flood peak attenuation may be less significant once wetland acreage in a watershed is greater than 10 percent (Ammon et al., 1981).

Methods: Select a watershed where the wetland acreage currently is less than 5 percent of the watershed size in humid regions and 3 percent in dry regions. The watershed, in this case, is the catchment area above the wetland replacement site, that is, the drainage area which flows into the wetland.

Included in this calculation of wetland size should be all other hydrologically connected existing wetlands and ponds.

Section 6.6
Water/Vegetation Proportions and Interspersion

Type: Site design

System: Nontidal riverine, lacustrine, palustrine

Criterion: In headwater areas in particular, create a wetland with a high proportion of vegetation in dense stands with little interspersed open water.

Rationale: Wetlands with relatively low proportions of open water to vegetation and low interspersion of water and vegetation are more capable of altering floodflows.

Vegetation slows floodwaters by creating frictional drag in proportion to stem density. Wetlands with dense stands of vegetation and with little open water are more capable of slowing flood water than open water alone. Channel roughness and thus the ability to retain floodwater increases with increasing vegetation density.

Methods: Create a wetland with a high proportion of vegetation coverage. Plant a concentration of woody plants to create high stem density.

Section 6.7
Vegetation Class

Type: Site design

System: Nontidal riverine, lacustrine, palustrine

Criterion: Create a wetland with forested or scrub/shrub vegetation.

Rationale: Vegetational resistance rapidly diminishes as the water depth becomes greater than the height of the vegetation (Camfield, 1977).

Methods: If the wetland is subject to overflow from a channel, the dominant vegetation class in the wetland should be forested or scrub/shrub.

Plant the wetland so the dominant vegetation class is forested or scrub/shrub. Conduct a literature search for tree and shrub species which are tolerant of frequently flooded conditions. In addition, ensure that the selected species are commercially available and are native to the region. Contact the local U.S.D.A. Soil Conservation Service District, the U.S. Fish and Wildlife Service, as well as local arboretums and nurseries for information on commercially available native plants. A partial listing of wetland and native plant suppliers may be found in Appendix B.

Section 6.8
Sheet Flow

Type: Site design

System: Nontidal riverine, lacustrine, palustrine

Criterion: Create a wetland where water flows primarily as sheet flow.

Rationale: Sheet flow, rather than channel flow, offers greater frictional resistance. Therefore, the potential for desynchronization of floodflows is greater when water flows through the wetland as sheet flow.

Methods: Create a wetland with a low gradient where water will move as sheet flow rather than in a channel. A wetland whose basin morphology allows for water to spread out rather than remain in a channel is preferable. A topographically flat wetland will lend itself to sheet flow conditions. Generally, gradients which adhere to those shown in Table 2, (Section 3.5, Nutrient Removal/ Transformation) will produce sheet flow conditions.

Velocity conditions less than 0.3 feet/second are generally recommended for wetlands greater than 20 feet wide.

Notes: It is critical that a careful examination be made of the watershed characteristics in order to safely alter floodflows. Acceptable storage times, volumes, and release rates must be calculated so that upstream and downstream conditions are not aggravated.

Chapter 7
Ground Water Recharge

Description of the Function:

Wetlands functioning to recharge ground water do so by holding surface water long enough to allow the water to percolate into the underlying sediments and/or bedrock aquifers.

To provide ground water recharge, the wetland's water source must come from surface water (channel flow or overland flow). Once the water reaches the wetland, it must be detained in the wetland basin. Porous underlying substrates (soil and/or bedrock) in turn allow water to pass through the substrate into the ground water system. Once the water reaches the ground water system, it aids in augmenting low flow of surface water streams and lakes. If it recharges a deep ground water system, water contributed by the wetland recharge may be used by public or private water supply systems.

General Design Concepts:

Design of a wetland for ground water recharge requires the expertise of a qualified hydrogeologist, since the potential ability of a wetland to recharge ground water is highly site specific. The hydrogeologist will assist in determining (1) the location of the underlying ground water table (piezometric surface) using sealed wells, and (2) evaluating the porosity of the soil and bedrock for water transmissibility to the underlying ground water system. Generally, appropriate soil and bedrock site conditions must be met if a wetland is to function adequately in ground water recharge. Therefore, one of the most critical criteria related to ground water recharge is site selection.

The movement of water from a wetland into the ground water is a function of the elevation head, or the elevation of the wetland in relation to the underlying ground water. The elevation head is increased where water is detained in the basin and discharged through a constricted outlet (Section 7.8), and where local topography is steeply sloping below the wetland (Section 7.9). Generally, these conditions can only exist in palustrine, lacustrine, and riverine systems (Section 7.1). Tidal, marine, and estuarine systems have little recharge potential because the elevation head is so low.

The porosity of the underlying soil and bedrock conditions determine the potential hydraulic conductivity of a wetland, or the ability of the water to flow from the wetland into the underlying strata (Sections 7.4 and 7.5).

In order for a wetland to contribute to the ground water system, its primary source of water must be from a surface water source including channel flow, precipitation, and upland runoff (Section 7.6). Wetlands receiving ground water as a water source are functioning as ground water discharge areas.

Specific Site Selection and Site Design Features:

Table 6 describes the features used in the ground water recharge function. With the exception of Section 7.6, Water Source, these features were derived from the WET predictors. All the features may be applied to nontidal riverine, lacustrine, and palustrine wetland systems.

Table 6. Ground water recharge site selection and site design features.

Section	Feature [1]	Site Selection	Site Design	Importance to Function [2]	Notes [3]
7.1	Wetland System	X		High	10
7.2	Land Cover of the Watershed	X		Moderate	21
7.3	Watershed Soils	X		High	24.4 renamed
7.4	Underlying Soils	X	X	High	24 renamed
7.5	Underlying Strata	X		Moderate	62
7.6	Water Source	X	X	High	new
7.7	Artificial Drainage Features		X	High	23 renamed
7.8	Outlet Characteristics		X	Moderate	8
7.9	Local Topography	X		Moderate	6
7.10	Water Chemistry	X		Moderate	new

1 Each feature applies to riverine, lacustrine, and palustrine wetland systems.

2 These ratings are generally derived from Volume I of WET (FHWA-IP-88-029). Some values were modified in relation to wetlands replacement criteria by Paul Adamus (U.S. EPA Environmental Research Lab, Corvallis, Oregon).

3 This column describes the WET 2.0 predictors used in deriving each feature. If only a number is given, it refers to the predictor number in WET 2.0. RENAMED means that the title of the WET 2.0 predictor was modified to show the new feature. NEW means the feature is not based on any given WET 2.0 predictor, but was inferred from other information.

WET 2.0 Predictors Not Used
1 Climate; 11 Fringe or Island; 32 Hydroperiod; 33 Most Permanent Hydroperiod; 34 Water Level Control; 35 Flooding Extent and Duration; 54 Ground Water Measurements; 59 Water Quality Anomalies; 60 Water Temperature Anomalies; 63 Discharge Differential

Section 7.1
Wetland System

Type:	Site selection
System:	Nontidal riverine, lacustrine, palustrine
Criterion:	Avoid selecting an estuarine wetland system or a tidally influenced system.
Rationale:	Riverine, lacustrine, and palustrine systems are more likely to perform ground water recharge functions. Estuarine or tidal systems have insufficient vertical head to recharge ground water sufficiently.
Methods:	Select a site which will not be tidally influenced.

Section 7.2
Land Cover of the Watershed

Type: Site selection

System: Nontidal riverine, lacustrine, palustrine

Criterion: Select a site where impervious surfaces dominate the watershed.

Rationale: If extensive paved surfaces surround the wetland, they may lower the ground water levels (piezometric contours), enhancing the potential for ground water recharge within the wetland itself.

Methods: Determine the amount of the watershed which contains impervious surfaces. These areas include streets, parking lots, and small lot development of less than 1/4 acre.

Compare this to the amount of pervious cover types (forest, agriculture, grasslands/pastures, and exposed soil areas) in the watershed to determine the dominant land cover type.

Section 7.3
Watershed Soils

Type: Site selection

System: Nontidal riverine, lacustrine, palustrine

Criterion: Select a replacement site in a watershed dominated by soils having a slow infiltration rate.

Rationale: Wetlands located in watersheds dominated by soils with slow infiltration rates are more likely to recharge ground water. Slow infiltration by watershed soils results in the delivery of more runoff to the wetland, making more water available for recharge by the wetland.

Methods: Evaluate the county soil survey for the replacement site to determine if one of the following conditions exist:

- The watershed is dominated by soils having a slow infiltration rate.
- The watershed is dominated by soils that are impermeable due to the texture, impeding layers (fragipan, duripan, claypan), high water table, shallow depth to unfractured bedrock, or soils that are frozen during the time of greatest seasonal runoff.

Section 7.4
Underlying Soils

Type: Site selection and site design

System: Nontidal riverine, lacustrine, palustrine

Criterion: Select a replacement site with permeable soils or create a substrate of permeable soils.

Rationale: Wetlands underlain by permeable soils with high infiltration rates are more likely to recharge ground water.

Methods: Select a site which contains permeable soils. Or, use permeable soils obtained from an off-site location for the replacement site's substrate. Soils to be used as a wetland substrate can be evaluated using the county soil survey, which describes the infiltration rate in inches per hour for each soil series. The soil infiltration rate should be at least 0.5 inches per hour to be considered permeable.

Coarse soils (sands and gravels) are highly porous and allow for rapid water percolation but may be too porous to support wetland conditions except in alluvial situations. On the other extreme, organic soils and clays may not transmit water rapidly enough. A soil type which is intermediate in porosity is therefore recommended, such as a sandy loam (Figure 14).

Notes: This evaluation should be combined with Section 7.5, Underlying Strata, which discusses underlying sediments and rock features.

(1.) Undesirable:
Recharge is occurring, but substrate is too porous to maintain wetland conditions.

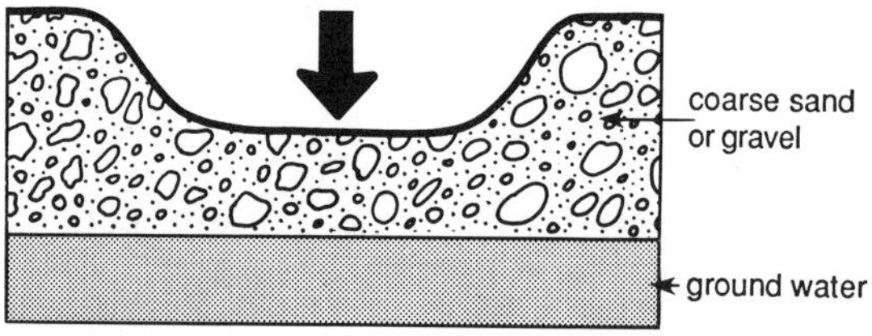

(2.) Undesirable:
Wetland conditions present, but the substrate seals the basin, preventing the water from percolating to ground water.

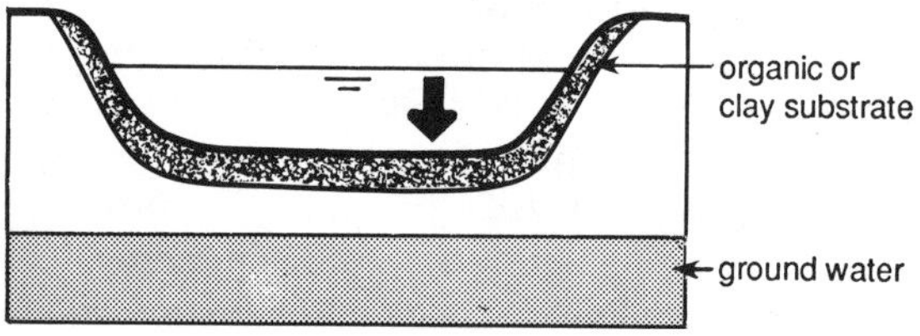

(3.) Desirable:
Substrate holds water long enough to promote wetland conditions, and is also permeable enough to allow ground water to recharge.

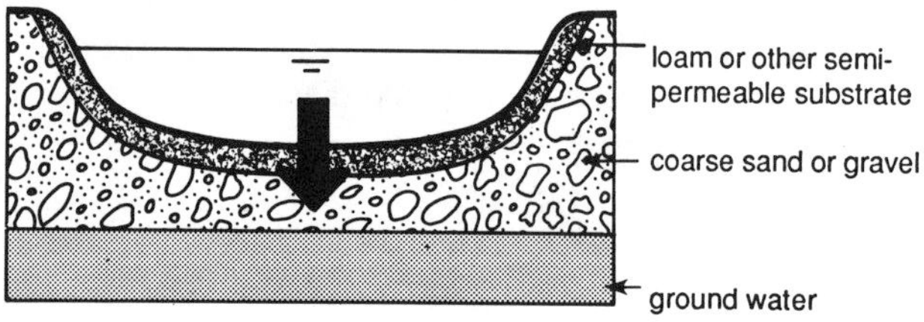

Figure 14. Underlying wetland soils for ground water recharge.

Section 7.5
Underlying Strata

Type: Site selection

System: Nontidal riverine, lacustrine, palustrine

Criterion: Select a replacement site underlain by thick, unstratified porous materials which are likely to recharge ground water.

Rationale: Porous sediments move water from a wetland to an aquifer. The presence of material characterized by low permeability can retard recharge.

Methods: Obtain any published ground water or geological surveys and data of the replacement area. Evaluate the porosity of the upper 50 feet of bedrock at the replacement site to determine if pervious sediments or highly fractured rock allowing ground water recharge are present. A minimum of 10 feet of porous material or highly fractured rock should be present below the wetland replacement site for significant ground water recharge to occur.

Because of its technical nature, this feature is best evaluated by a qualified hydrogeologist.

Notes: This evaluation should be combined with Section 7.4, Underlying Soils, which discusses soil conditions of the wetland substrate.

Section 7.6
Water Source

Type: Site selection and site design

System: Nontidal riverine, lacustrine, palustrine

Criterion: Use a surface water source for the wetland's principal water supply.

Rationale: If the water supply for a wetland originates from a surface water source, such as channel flow, overland flow, and precipitation, the wetland will act to add this water to the ground water system. On the contrary, wetlands whose principal water supply is contributed by ground water are acting as discharge areas, not recharge areas.

Methods: For methods, see Section 3.2, Nutrient Removal/Transformation.

Section 7.7
Artificial Drainage Features

Type: Site design

System: Nontidal riverine, lacustrine, palustrine

Criterion: Avoid artificial drainage features.

Rationale: Drainage structures which cause water to flow out of the wetland lower the hydraulic gradient.

Methods: Avoid the use of any features which would allow the water to rapidly flow out of the wetland. This includes channels, levees, ditches, canals, or similar types of drainage features.

Section 7.8
Outlet Characteristics

Type: Site design

System: Nontidal riverine, lacustrine, palustrine

Criterion: Create a constricted surface water outlet or no outlet at all.

Rationale: Wetlands with a constricted outlet or no outlet usually have a hydraulic gradient favoring ground water recharge. Also, the longer the detention time of the water in the wetland, the greater the opportunity for the water to percolate into the underlying substrate.

Wetlands with constricted outlets or no outlet generally have fluctuating water levels, which will periodically inundate adjacent unsaturated soils. Soils which are seasonally or temporarily flooded are more likely to transmit water than saturated soils.

Methods: For wetlands with channel flow, a constricted outlet should be less than 1/3 the average width of the wetland at annual high water or the cross sectional area of the outlet should be less than 1/3 the cross sectional area of the inlet.

For wetlands with no channel flow, the width of the outlet should be less than 1/10 the average width of the wetland. Also, see Section 3.6, Nutrient Removal/Transformation, for methods.

Notes: Annual high water is specifically defined as the two year flood.

Section 7.9
Local Topography

Type: Site selection

System: Nontidal riverine, lacustrine, palustrine

Criterion: Select a site where the topography and the ground water table slope sharply downward from the wetland.

Rationale: The slope of the ground water table often parallels the topography of the land surface. When local topography slopes sharply away from a wetland, the water table will also slope away, resulting in a hydraulic gradient favorable for the movement of water from the wetland into the ground water.

Methods: The location of the ground water surface in relation to the wetland replacement site should be determined by a qualified hydrogeologist. Sealed wells, or piezometers, can be used to determine the water table location and elevation (piezometric surface).

Generally, a site favoring ground water recharge will also meet one of the following conditions (Figure 15):

- The drop in elevation from the replacement site to a point two miles downslope is greater than the drop in elevation from a point two miles upslope.
- The replacement site is located within two miles of a topographic divide between two major watersheds. (A major watershed contains a river channel of at least 100 feet in width from bank to bank).

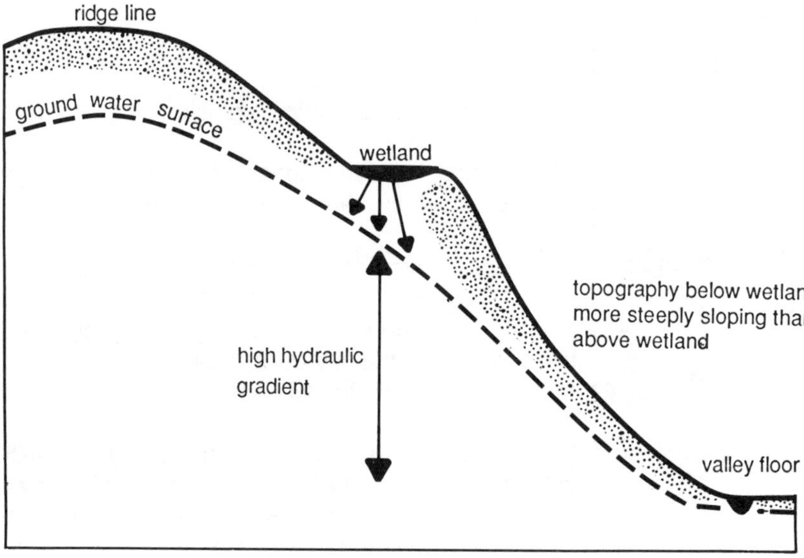

Figure 15. Topography with a hydraulic gradient favoring ground water recharge.

Section 7.10
Water Chemistry

Type: Site selection

System: Nontidal riverine, lacustrine, palustrine

Criterion: Avoid using contaminated surface water for ground water recharge.

Rationale: Contaminated surface waters recharged to ground water could be a potential problem for public or private use.

Methods: Conduct water quality testing for the surface water source for the replacement site. Water quality for the site should meet those criteria and standards set forth by the state in which the site is located.

Chapter 8
Production Export

Description of the Function:

Production export is the production of organic material and its subsequent physical transport out of a wetland to areas downstream or to deeper waters within the same basin. This organic material is then added to the food chain, where it is eaten by primary consumers (fish and aquatic invertebrates). Wetlands are often referred to as one of the most productive biological systems because of their export of large amounts of organic material, and because they are often hydrologically linked to other systems.

The direct relationship between production export from a wetland to a specific consumer group within a defined downstream or deep water area is difficult to establish. This is principally because the dispersal of organic material from a wetland is not consistent in quantity or quality throughout the year, nor is all exported material cycled to one particular area. Additionally, the organic input to a given deep water or downstream area may come from multiple sources.

Because of this, replacement wetland design for production export is developed in this guidebook with the concept of maximizing production export out of the wetland, without a specific designated receiving "target" consumer or area. The target area must be identified on a project and site specific basis. Generally, the target site must be directly linked to the replacement wetland hydrologically.

General Design Concepts:

There are three principal aspects of a wetland which establish its ability to produce and export organic material:

(1) plant productivity,
(2) nitrogen fixing ability, and
(3) capacity for physical dispersal of food sources.

The productivity of wetland macrophytes (emergent and aquatic bed plants) is high, ranging up to 1980 g/m²/yr. Macrophytes further increase productivity by providing attachment surfaces for epiphytic algae, and intercepting floating detritus and microbes, causing them to be seasonally deposited in a wetland. However, plant forms other than macrophytes are also highly productive, including benthic and epiphytic algae, woody plants, and phytoplankton.

Although scrub/shrub and forested macrophytes produce less organic material than emergent and aquatic bed macrophytes, the larger macrophytes produce materials during periods when the smaller macrophytes are in a less productive state, therefore contributing seasonal stability to the food export supply (Section 8.1).

Carbon is contributed to a wetland system by all plants, but nitrogen, another critical element, is added only by nitrogen fixing bacteria and some blue-green algae. Although bacteria and algae are commonly found in floating or benthic mats, they are also found in association with certain macrophytes, which foster a symbiotic relationship. These species may be encouraged to thrive in a wetland replacement area (Section 8.4).

Transport of organic materials is the most critical element in a wetland's production export capability, regardless of the volume of organic material produced. A wetland design should therefore concentrate on the physical dispersal of organic materials out of the wetland to adjacent downstream or deep water areas.

To do this, the wetland must be linked hydrologically to downstream wetlands via an outlet, or operate as an adjacent fringe wetland in an existing basin (Sections 8.5, 8.15, and 8.16). Major export occurs during flood events, since the energy produced by these events picks up and carries organic matter which has accumulated on wetland substrates (Sections 8.6 and 8.10). However, wetlands which are permanently flooded may also function well in production export, as long as the contact between surface water and macrophytes is maximized.

Generally, surface water must flow through a wetland rather than stagnate (Section 8.8). In addition, primary productivity and decomposition rates are higher in wetlands with flowing water and sheet flow (Section 8.7). However, high water velocities discourage plant growth.

The balance between vegetation and open water is important, as the vegetation must not be so dense as to impede circulation, but have a density high enough to be productive (Sections 8.2 and 8.3). The optimum condition therefore is a high density of macrophytes in small patches throughout the wetland.

Specific Site Selection and Site Design Features:

The following table describes the features used in the production export function. With the exception of Section 8.15, Target Area, these features were derived from the WET predictors. The features may apply to some or all of the wetland systems.

Table 7. Production export site selection and site design features.

Section	Feature	System [1]	Site Selection	Site Design	Importance to Function [2]	Notes [3]
8.1	Vegetation Class and Form Richness	E,R,L,P		X	Moderate	12, 17 renamed
8.2	Water/Vegetation Proportions and Interspersion	E,L,P		X	Moderate	15.1, 31 renamed
8.3	Vegetated Width	R		X	Moderate	36
8.4	High Plant Productivity	E,R,L,P		X	Moderate/ High	51
8.5	Outlet Characteristics	E,R,L,P		X	High	8 renamed
8.6	Flooding Extent and Duration	E,R,L,P	X	X	Moderate/ High	35
8.7	Sheet Flow	E,R,L,P		X	High	15.2 renamed
8.8	Channel Gradient and Water Velocity	E,R,L,P		X	Moderate/ High	7,41 renamed
8.9	Fetch/Exposure	E,R,L,P	X		Moderate	19
8.10	Water Level Control	E,R,L,P	X		Moderate	34
8.11	Wetland/ Watershed Ratio	E,R,L,P	X		Low	5
8.12	Watershed Size	E,R,L,P	X		Moderate	4
8.13	Substrate Type	R,L,P	X	X	Moderate/ High	45
8.14	pH	E,R,L,P	X		Moderate	47
8.15	Target Area	E,R,L,P	X		High	new
8.16	Fringe/Island Wetlands	P		X	Moderate	11, 14 renamed

1 E - estuarine, R - riverine, L - lacustrine, P - palustrine

2 These ratings are generally derived from Volume I of WET (FHWA-IP-88-029). Some values were modified in wetlands replacement criteria by Paul Adamus (U.S. EPA Environmental Research Lab, Corvallis, Oregon).

3 This column describes the WET 2.0 predictors used in deriving each feature. If only a number is given, it refers to the predictor number in WET 2.0. RENAMED means that the title of the WET 2.0 predictor was modified to show the new feature. NEW means the feature is not based on any given WET 2.0 predictor, but was inferred from other information.

WET 2.0 Predictors Not Used
1 Climate; 2 Acreage; 10 Wetland System; 13 Vegetation Class; 22 Flow, Deposition, Gradient; 28 Direct Alteration; 55 Suspended Solids; 56 Dissoloved Solids or Alkalinity; 57 Eutrophic Condition

Section 8.1
Vegetation Class and Form Richness

Type: Site design

System: Estuarine, riverine, lacustrine, palustrine

Criterion: Include a vegetative cover which is predominantly aquatic bed and emergent species. Emphasize high vegetation form richness.

Rationale: Aquatic bed species are generally the most productive species, since they can transfer nutrients from the sediment to the water column, and decompose more rapidly than other vegetation forms. However, emergent vegetation is also highly productive, more so than woody vegetation.

The wetland, however, should support each of the above mentioned vegetation classes (aquatic bed, emergent, scrub/shrub, and forested), since plant detritus is decomposed and transported at different species-specific rates. A variety of vegetation classes will therefore make production export rates more balanced throughout the year.

Methods: Plant the wetland so it is dominated by aquatic bed and emergent species, but also contains scrub/shrub and forested classes in the design.

Conduct a literature search for commercially available species which are native to the region. Local nurseries and arboretums as well as the local district office of the U.S.D.A. Soil Conservation Service may provide information on availability of native species.

A partial listing of wetland and native plant suppliers may be found in Appendix B.

Section 8.2
Water/Vegetation Proportions and Interspersion

Type:	Site design
System:	Estuarine, lacustrine, palustrine
Criterion:	Intersperse water and vegetation. A portion of the wetland should remain unvegetated, open water.
Rationale:	Vegetation which is too dense may inhibit water circulation, thereby inhibiting the export of nutrients out of the wetland. On the other extreme, vegetation densities which are too sparse are not productive, and may be incapable of retaining drifting organic material, an important food source for invertebrates.
Methods:	Design bottom contours that will support a mosaic of small patches of dense vegetation interspersed with open water (Figure 16). Height and density of the vegetation should not completely shade the open water areas since shading reduces the productivity of benthic algae. Avoid vegetation densities which impede water circulation.
Notes:	If the system is riverine, Section 8.3, Vegetated Width, may be used in place of this feature.

Create a balance of vegetated and open water areas which are well interspersed.

Figure 16. Example of a wetland with well balanced proportions and a high interspersion of vegetation and open water.

Section 8.3
Vegetated Width

Type: Site design

System: Riverine

Criterion: Create a wetland whereby the average width of wetland vegetation perpendicular to the channel flow is greater than 20 feet.

Rationale: The wider the vegetated width, the greater the potential for production export.

Methods: Establish a predominance of aquatic bed and emergent vegetation adjacent to the river channel, with widths exceeding 20 feet. Woody vegetation (scrub/shrub and forest) is less productive, but still desirable (see Section 8.1, Vegetation Class and Vegetation Form Richness).

A partial listing of wetland and native plant suppliers may be found in Appendix B.

Notes: Section 8.2, Water/Vegetation Proportions and Interspersion, is not necessary if this feature is used.

Section 8.4
High Plant Productivity

Type: Site design

System: Estuarine, riverine, lacustrine, palustrine

Criterion: Select wetland plants which are highly productive.

Rationale: The higher the productivity, the greater the amount of organic matter potentially available for export.

Methods: Conduct a literature search for plant species which are high in annual primary productivity for the particular region of the replacement site. A partial list of highly productive plants includes the following (Adamus, 1983):

Aquatic Bed:

yellow cow lily	*Nuphar luteum*
mosquito fern and duckweed	*Azolla mexicana* and *Spriodela polyrhiza*
Chara	*Chara spp.*

Emergent:

salt meadow rush	*Juncus gerardi*
switchgrass	*Panicum virgatum*
salt meadow cordgrass	*Spartina patens*
salt marsh cordgrass	*Spartina alterniflora*
big cordgrass	*Spartina cynosuroides*
broad-leaf cattail	*Typha latifolia*
narrow-leaf cattail	*Typha angustifolia*
sedge species	*Carex spp.*

Nitrogen Fixing Wetland Host Plants:

sweet gale	*Myrica gale*
speckled alder	*Alnus rugosa*
northern bayberry	*Myrica pensylvanica*
rush species	*Juncus spp.*

Nitrogen fixing wetland host plants support a symbiotic relationship with nitrogen fixing blue-green algae.

In general, aquatic bed and emergent plants have the highest annual net primary productivity. These classes can therefore be used extensively in wetland replacement areas when production export is the goal of the functional replacement (see Section 8.1).

Wetland morphology is a major factor which determines the ability of macrophytes to exist. The wetland must be shallow, sheltered, soft-bottomed, and unshaded to maximize macrophyte growth.

Section 8.5
Outlet Characteristics

Type: Site design

System: Estuarine, riverine, lacustrine, palustrine

Criterion: Create a permanent or intermittent outlet, and a permanent or intermittent inlet.

Rationale: Organic materials are transported out of a wetland when there is an outlet connecting it with other wetlands. A defined inlet also allows for better flushing of the wetland.

Methods: Include a defined inlet and outlet which are hydrologically connected to existing streams or wetlands.

Notes: This feature should be used in conjunction with Section 8.15, Target Area.

Section 8.6
Flooding Extent and Duration

Type: Site selection and site design

System: Estuarine, riverine, lacustrine, palustrine

Criterion: Create seasonally flooded conditions. Or, locate the replacement site adjacent to surface water which is subject to seasonal flooding or experiences a seasonal high water table.

Rationale: Seasonal flooding is the optimal hydroperiod regime for the export of detritus. The drier periods enhance decomposition, while the flooded periods export the decomposed detritus. Semipermanently and permanently flooded hydroperiods are desirable, but to a lesser degree.

Methods: Select a replacement site which is subject to or is adjacent to an area which is subject to a seasonal high water table or seasonal flooding. Create a hydroperiod which allows for seasonal flooding.

For site selection, select a site whose seasonal high water table is close to the ground surface.

For site design, establish the elevation at the site for permanently flooded conditions and for seasonally flooded conditions (see Chapter 2, Wetland Hydrology, for methods). It is important to establish the number of days seasonally flooded conditions will be present in an average year (Figure 17(a)).

Second, select the basin size and surface area for the permanently flooded zone of the wetland (see Figure 17(b)).

Consulting Figure 11, determine the surface area required for the wetland's seasonally flooded zone to be considered significantly expanded (Figure 17(c)). This will appear as a "basin within a basin" in cross section, when the seasonally flooded and permanently flooded elevations are tied into the original ground elevation (Figure 17(d)).

(a) Determine elevations

(b) Size basin for permanently flooded zone

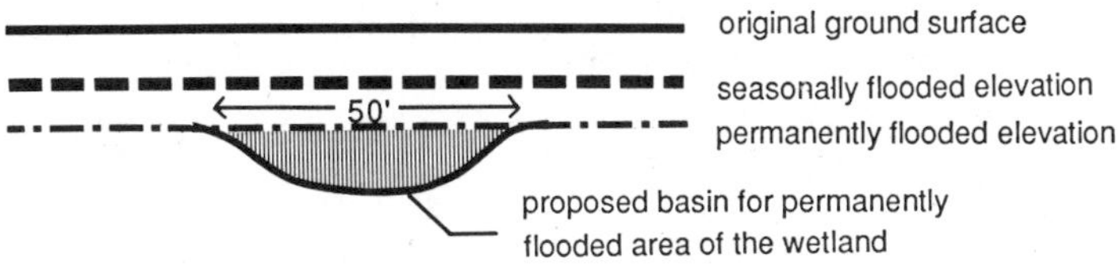

Example: Surface area of the permanently flooded zone is 50%.

(c) Size basin for seasonally flooded zone

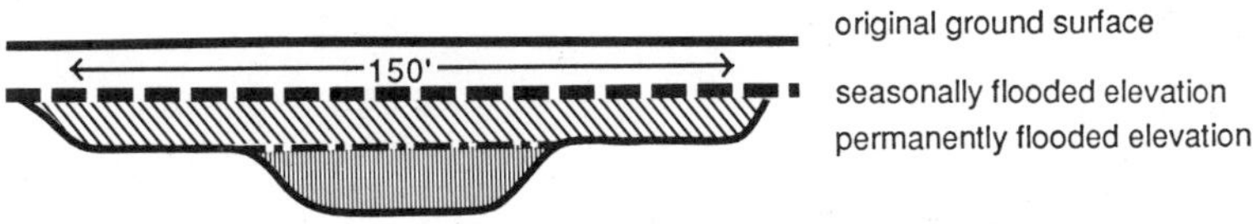

Example: Surface area of the seasonally flooded zone is 200% of the surface area of the permanently flooded zone, which meets the criterion of significantly expanded surface area as defined in Figure 11.

(d) Tie in elevations with original ground

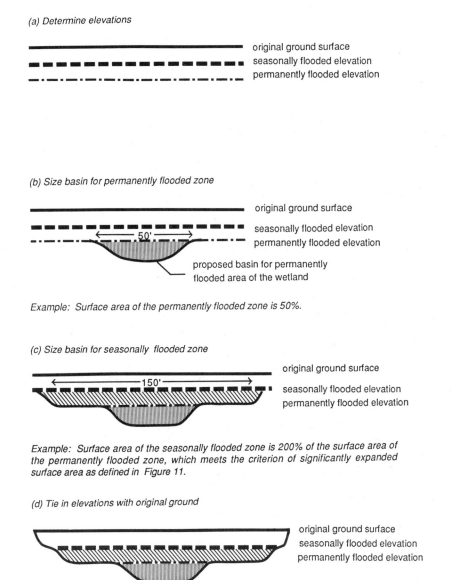

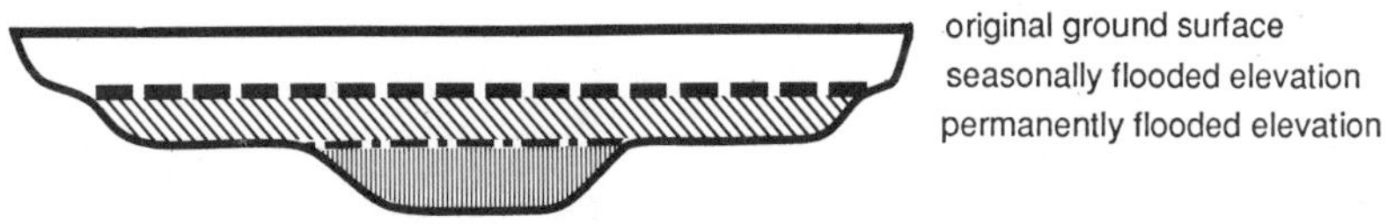

Figure 17. Sizing a wetland basin for significantly expanded seasonal flooding (ground water system).

For site selection, locate a site which experiences a significant rise in flood level on a seasonal basis.

For site design, consult Chapter 2, Wetland Hydrology, for wetlands using a surface water supply. The cross sectional area of the stream and its associated wetland must accommodate flood waters for the specified extent and duration identified in Figure 11.

Section 8.7
Sheet Flow

Type: Site design

System: Estuarine, riverine, lacustrine, palustrine

Criterion: Create sheet flow rather than channel flow.

Rationale: The greater the duration of contact between vegetation and flowing water, the greater the potential for production export.

Methods: Flow through the wetland should not be confined to channels. Rather, water entering the wetland should be encouraged to flow in contact with wetland vegetation to the extent possible. To do this, the gradient of the wetland should be minimized.

Notes: This feature may conflict with Section 8.8, Channel Gradient and Water Velocity. Either create sheet flow conditions or provide non-depositional velocities.

Section 8.8
Channel Gradient and Water Velocity

Type: Site design

System: Estuarine, riverine, lacustrine, palustrine

Criterion: Create a gradient whereby water flows through the wetland rather than stagnates. Or, create moderate peak annual flow velocities.

Rationale: The potential for production export increases with increasing flow velocities to an extent. Primary productivity and decomposition rates are usually higher in wetlands with flowing water.

Methods: The gradient of the wetland should be steep enough to create non-depositional conditions.

Use Table 8 to determine the minimum slope (gradient) of the wetland channel for maintaining sheet flow. Velocities should range between 0.3 to 1.5 feet/second during peak annual flow.

Table 8. Gradient necessary to create non-depositional velocity conditions.*

Mean Depth(ft)	Densely Wooded Floodplains	Densely Vegetated Emergent Wetlands not Totally Submerged by Floodflow	Moderately Vegetated or Totally Submerged Emergent Wetlands, or with Boulders	Unobstructed Channels
<0.5	>0.0250	>0.0100	>0.0038	>0.0018
0.5 - 1	>0.0150	>0.0060	>0.0023	>0.0012
1 - 2	-----------	>0.0030	>0.0012	>0.0006
2 - 3	-----------	>0.0017	>0.0006	>0.0003
3 - 4**	-----------	>0.0013	>0.0005	>0.0002
4 - 6**	-----------	>0.0008	>0.0003	>0.0001
6 - 8**	-----------	>0.0006	>0.0002	>0.0001
8 - 10**	-----------	>0.0004	>0.0002	----------
10 - 12**	-----------	>0.0003	>0.0001	----------

* Interpreted form SCS curves for channel flow and Manning's roughness coefficients.
** Assumes width, perpendicular to flow is less than 8 feet. If channel is 8 - 20 feet wide, the value in the row immediately below the value identified should be used.

Source: Adamus et al., 1987

Section 8.9
Fetch/Exposure

Type:	Site selection
System:	Estuarine, riverine, lacustrine, palustrine
Criterion:	Locate the wetland in a moderately sheltered area where exposure to wind and waves is moderate.
Rationale:	Some wind and wave exposure favors the export of organic materials and associated nutrients. Both wind and waves cause vertical mixing of the water column, suspending organic materials, thus making them more susceptible to transport out of the wetland.
Methods:	Locate the wetland replacement site in an area which is not completely wind-buffered by vegetation or topographic relief.

Section 8.10
Water Level Control

Type: Site selection

System: Estuarine, riverine, lacustrine, palustrine

Criterion: Locate the wetland in an area where the water levels will not be controlled artificially upstream.

Rationale: When flooding frequency is reduced by dams or other artificial control structures upstream of a wetland, the potential for production export from the wetland decreases. Water control structures which lower stream discharge reduce the flooding frequency of downstream wetlands, thereby decreasing production export. In addition, reduced flow through a wetland can decrease productivity and kill vegetation.

Methods: Select a replacement area where the surface water is not controlled by any upstream structures. Use U.S.G.S. data, aerial photographs, and other regional mapping to determine if any dams or other artificial control structures are located upstream from the replacement area.

Section 8.11
Wetland/Watershed Ratio

Type: Site selection

System: Estuarine, riverine, lacustrine, palustrine

Criterion: Locate the wetland so that it occupies a minimum of 20 percent of its drainage area.

Rationale: Larger wetlands are potentially greater exporters of nutrients than smaller wetlands.

Methods: Select a subwatershed whose size enables the replacement wetland to comprise a minimum of 20 percent of the area.

Notes: If this feature is used in the site selection process, delete consideration of Section 8.12, Watershed Size. This feature is most applicable to small watersheds, whereas Section 8.12 applies to larger watersheds.

Section 8.12
Watershed Size

Type: Site selection

System: Estuarine, riverine, lacustrine, palustrine

Criterion: *For estuarine, lacustrine, and palustrine:*
Select a site whose watershed is greater than one square mile.

For riverine:
Locate the wetland in a large watershed, at least a fifth order stream.

Rationale: Wetlands with large watersheds are more likely to be flushed by flood waters because of greater runoff. Flushing by flood waters will ensure production export.

Methods: *For estuarine, lacustrine, and palustrine:*
Determine the size of the watershed of the proposed replacement wetland site. Watersheds greater than one square mile are the most desirable.

For riverine:
Select a watershed for the wetland greater than 100 square miles in size. Also, the higher the order the stream, the more preferable. At least a fifth order stream is recommended.

Notes: If this feature is used in the site selection process, delete consideration of Section 8.11, Wetland/Watershed Ratio. This feature is most applicable to large watersheds, whereas Section 8.11 applies to smaller watersheds.

Section 8.13
Substrate Type

Type: Site selection and site design

System: Riverine, lacustrine, palustrine

Criterion: Avoid locating the replacement wetland in an area which has a sand substrate, or use a substrate material other than sand.

Rationale: Sand can be an unstable substrate and contain few nutrients, making plant productivity low.

Methods: Select a site which does not have sand as a substrate, or design the wetland to use any type of substrate except sand.

Notes: Sand may be a suitable substrate in estuarine systems where nutrients necessary for plant growth are likely to be present in the water.

Section 8.14
pH

Type:	Site selection
System:	Estuarine, riverine, lacustrine, palustrine
Criterion:	Select a location with circumneutral water having pH between 6.0 to 8.5.
Rationale:	Productivity of wetland plants is highest where pH is neutral. Higher productivity in turn provides greater volumes of material for export.
Methods:	Select a site which will support a circumneutral water environment. Sampling the proposed site for pH will provide information needed for this feature.

Section 8.15
Target Area

Type:	Site selection
System:	Estuarine, riverine, lacustrine, palustrine
Criterion:	Locate the replacement wetland so it is physically connected to downstream wetlands or deepwater areas which serve as target areas. Target areas are generally aquatic systems which support high densities of important fish and invertebrate species, particularly nursery and spawning sites.
Rationale:	Target areas, or the receiving consumer of the organic matter, must be hydrologically linked to the replacement wetland. Target areas receive the organic by-products of the highly productive upstream replacement wetland.
Methods:	Identify specific target areas located downstream from the replacement wetland.

Section 8.16
Fringe/Island Wetlands

Type: Site design

System: Palustrine

Criterion: Create an island or fringe wetland adjacent to standing or flowing water.

Rationale: In fringe or island wetlands, organic materials are readily transported downstream.

Methods: Create an island wetland within a flowing or standing body of water, or design a fringe wetland along a channel. A fringe wetland should have a width less than three times the width of the adjacent channel. An island wetland should have a surface area less than one-third the area of an adjacent standing body of water. (See Glossary for definition of fringe and island wetlands.)

Chapter 9
Aquatic Diversity/Abundance

Description of the Function:

This function refers to the support of diverse and/or abundant fish and invertebrates. The ability of a wetland to fulfill this function is dependent upon the type and quality of habitat it provides. Several critical habitat factors affecting aquatic diversity and abundance are salinity, temperature, substrate, current velocity, dissolved oxygen, vegetation composition, and the interspersion of vegetation and water.

Nearly all freshwater and many saltwater fisheries, at some stage in their life cycle, require shallow water areas which wetlands provide. However, not all wetlands can serve in this capacity. Many noncontiguous wetlands are too susceptible to lethal stagnation conditions. These conditions cause low dissolved oxygen levels and high temperatures, and will not support a sustained fishery. Many acidic moss wetlands, hypersaline wetlands, and very shallow wetlands are also unfavorable.

Wetland vegetation provides a source of nutrients, protective cover, and temperature moderation by providing shade. Vegetation which is attractive to aquatic invertebrates will aid in sustaining an abundance of fish species dependent upon those invertebrates. Vegetation, in the form of detritus, is critical to the fishery food chain in most riverine systems (Marzolf, 1978). As a directly consumed food source, algae and phytoplankton are more usable to most aquatic consumers than tissues of aquatic or emergent plants. Some wetlands are important to fisheries for their ability to recycle and export nutrients rather than providing adequate habitat for organisms within the producing wetland. Partially submerged wetland plants provide optimal environments for most aquatic invertebrate and juvenile fish communities (Turner, 1977; Boesch and Turner, 1984). Cover can also be provided by overhanging trees, logs, boulders, bottom sediments, undercut banks, water turbulence, and turbidity.

Although a diversity of vegetation in a wetland is needed in turn to support a diversity of invertebrates and fish, vegetation can be detrimental if it is too dense. Channels, pools, or other open water areas are needed for fish movement. Densely vegetated wetlands also have fewer large fish, a lower invertebrate density (Swanson and Meyer, 1979), and perhaps reduced fish population.

General Design Concepts:

In that aquatic organisms have diverse habitat needs, numerous factors require consideration. Factors to be evaluated in the development of a wetland site for aquatic diversity and abundance include water quality, water quantity, cover, substrate, interspersion, and food. The key site selection and site design concept for this function is diversity of the habitat conditions themselves, assuming that a diversity of conditions will in turn create a diversity of organisms.

Because of this, the first step in developing a wetland design is to evaluate the replacement site in relation to other existing wetlands which are hydrologically linked to the replacement site (Sections 9.2 and 9.14). A replacement wetland is most valuable if it increases the aquatic diversity of the existing wetland context. Therefore, it is essential to determine the existing habitat conditions and limiting factors of the hydrologic wetland system in the context of the replacement site.

Temperature, in an aquatic environment, is the most easily controlled water quality factor. High temperatures, a limiting factor for many aquatic organisms, can be controlled by providing shade from overhanging vegetation (Section 9.8), deep pools (Section 9.13), and flowing water (Section 9.5). Other chemical factors such as alkalinity, pH, and oxygen are more difficult to control within a wetland design and therefore need to be considered in site selection (Section 9.19). Suspended sediment and turbidity can be controlled to a degree by keeping the wetland gradient gradual (Section 9.6).

The hydrology of a wetland is another critical component in aquatic diversity, and includes depth, velocity, and hydroperiod. A diversity of depth ranges within a wetland or wetland system is desirable. Shallow water depths support vegetation growth which provides fish cover and habitat for aquatic invertebrates. Deep water is important for fish movement and refuge in drought periods (Section 9.1).

A range of water velocities is desirable for aquatic diversity. To avoid stagnation, water should flow rather than stand. Maximum velocities are approximately 1.5 feet/second, which enable small fish to maintain their body position in flowing water. Within this range a variation is desirable, in that the slower moving velocity areas support the growth of wetland vegetation and the faster areas provide clean spawning gravel areas and reduce plant growth which may inhibit fish movement.

Generally, a combination of hydroperiods is optimal for fish and invertebrate abundance. For riverine, palustrine, and lacustrine systems, seasonally or semipermanently flooded wetlands are generally most productive, but they must be hydrologically connected to a permanently flooded area which provides refuge in drought periods. For estuarine systems, regularly flooded or irregularly exposed hydroperiods support the highest diversity.

Fish cover may be provided by low overhanging trees, aquatic vegetation, logs, boulders, and sediments (Sections 9.11 and 9.12). Wetland plants, particularly aquatic bed and emergent species, provide optimal cover for most aquatic invertebrates and juvenile fish (Sections 9.7 and 9.12).

Aquatic plants are the base of the wetland food chain. Detritus from terrestrial, and especially aquatic plants, serves as the food base for benthic organisms and macroinvertebrates. A diversity of food sources best supports a diversity of aquatic organisms. To that end, several forms of vegetation (Sections 9.9, 9.10, and 9.12) in a wetland are desirable, as are a variety of physical habitat types (Section 9.13).

Access to wetland vegetation for fish must be maintained by providing unvegetated channels throughout the wetland (Section 9.7).

Specific Site Selection and Site Design Features:

Table 9 describes the features for aquatic diversity and abundance. The type of wetland system is a critical determinant of the design features which are suitable. Because of this, the appropriate systems are noted for each of the function features.

Table 9. Aquatic diversity/abundance site selection and site design features.

Section	Feature	System [1]	Site Selection	Site Design	Importance to Function [2]	Notes [3]
9.1	Hydroperiod	E,R,L,P	X	X	High	32, 33 renamed
9.2	Outlet Characteristics	L,P		X	Moderate	8 renamed
9.3	Flooding Extent and Duration	R,L,P	X	X	High	35
9.4	Water Source	R,L,P	X		Moderate	new
9.5	Water Level Control	E,R,L,P	X		Moderate/High	34
9.6	Channel Gradient and Water Velocity	R		X	Moderate	7, 41
9.7	Water/Vegetation Proportions	E,R,L,P		X	High	31
9.8	Vegetated Canopy	R	X	X	High	20
9.9	Vegetation Form Richness	L,P		X	High	17
9.10	Vegetation/Water Interspersion	E		X	High	15.1 renamed
9.11	Aquatic Habitat Features	R		X	Moderate	49
9.12	Vegetation Class	R,L,P		X	High	13
9.13	Physical Habitat Interspersion	L,P		X	Moderate	46
9.14	Diversity Enhancement	E,R,L,P	X	X	High	new
9.15	Watershed Size	R,L,P	X		Low	4.2 renamed
9.16	Substrate Type	L,P	X	X	Moderate	45
9.17	Wetland Acreage	L,P	X		Moderate	2 renamed
9.18	Sediment and Contaminant Sources	R,L,P	X		Low	25, 27
9.19	Water Chemistry	E,R,L,P	X		Moderate	new
9.20	Wetland/ Watershed Ratio	E	X		Low	5

1 E - estuarine, R - riverine, L - lacustrine, P - palustrine.

2 These ratings are generally derived from Volume I of WET (FHWA-IP-88-029). Some values were modified in relation to wetlands replacement criteria by Paul Adamus (U.S. EPA Environmental Research Lab, Corvallis, Oregon).

3 This column describes the WET 2.0 predictors used in deriving each feature. If only a number is given, it refers to the predictor number in WET 2.0. RENAMED means the title of the WET 2.0 predictor was modified to show the new feature. NEW means the feature is not based on any given WET 2.0 predictor, but was inferred from other information.

WET 2.0 Predictors Not Used
1 Climate; 10 Wetland System; 11 Fringe Wetland or Island; 12 Vegetation Class/Subclass (Primary); 15.2 Sheet Flow; 18 Upland/Wetland Edge; 21 Land Cover of the Watershed; 23 Ditches/Canals/Channelization/Levees; 28. Direct Alteration; 40 Temperature; 47 pH; 48 Salinity and Conductivity; 52 Freshwater Invertebrate Density; 53 Tidal Flat Invertebrate Density/Biomass; 55 Suspended Solids; 56 Dissoloved Solids; 57 Eutrophic Conditions; 61 D. O.

Section 9.1
Hydroperiod

Type: Site selection and site design

System: Estuarine, riverine, lacustrine, palustrine

Criterion: *For lacustrine and palustrine systems:*
Include one of the following hydroperiods for at least a portion of the wetland or contiguous wetland or deepwater:

- permanently flooded nontidal (surface water present throughout the year), and
- intermittently exposed nontidal (surface water present except drought years).

Or, select a site which will be hydrologically connected to a wetland with a permanently flooded or intermittently exposed hydroperiod.

The remaining portion should be seasonally or semipermanently flooded (see Section 9.3, Flooding Extent and Duration).

For estuarine systems:
Include a regularly flooded or irregularly exposed hydroperiod.

For riverine systems:
Select a hydroperiod whereby the majority of the wetland is seasonally flooded, with a portion of it permanently flooded or intermittently exposed. Or, select a site which is hydrologically connected to a wetland with a permanently flooded or intermittently exposed hydroperiod.

Rationale: A higher diversity of fish and invertebrates is present in wetlands which support at least some areas of permanent water or are hydrologically connected to permanent water. Permanent water provides a greater amount of aquatic habitat for longer periods of time and provides a refuge for aquatic life when other areas are dry.

Methods:

For palustrine and lacustrine systems:
Create a wetland with a minimum of 10 percent of its area in a permanently flooded area. Or, locate the wetland adjacent to an existing wetland with a permanently flooded hydroperiod and provide a surface water connection having a permanent depth of at least four inches to the replacement wetland.

For estuarine systems:
Ensure that the wetland will include regularly flooded (intertidal) or irregularly exposed hydroperiods.

For riverine systems:
Create a wetland with at least 10 percent of its area in a permanently flooded or intermittently exposed hydroperiod, or a combination of the two. Or, locate the wetland adjacent to an existing wetland with a permanently flooded or intermittently exposed hydroperiod, providing a surface water connection between the two. The remaining area should support a seasonally flooded and/or semipermanently flooded hydroperiod, or a combination of the two.

Chapter 2, Wetland Hydrology, describes specific methods for deriving each hydroperiod described above.

Notes:

This feature should be considered in conjunction with Section 9.4, Water Source.

Section 9.2
Outlet Characteristics

Type: Site design

System: Lacustrine, palustrine

Criterion: Provide both an inlet and an outlet for the wetland.

Rationale: A surface water connection to adjacent waters and wetlands enables access to the wetland by colonizing organisms, and allows the input of natural food material.

Methods: Connect the surface water inlet and outlet of the replacement wetland to adjacent streams and wetlands.

Section 9.3
Flooding Extent and Duration

Type: Site selection and site design

System: Riverine, lacustrine, palustrine

Criterion: Ensure that seasonal flood events will significantly expand the wetland's size for extended periods of time.

Select a site adjacent to surface water which is subject to seasonal flooding or a site which experiences a seasonal high water table.

Rationale: Spring floods provide essential spawning, feeding, and nursery areas for many invertebrate and fish species. Flooding also reduces the magnitude of limiting factors such as dissolved oxygen.

Methods: See Section 4.6, Sediment/Toxicant Retention, and Chapter 2, Wetland Hydrology, for determination of flooding extent and duration.

Section 9.4
Water Source

Type:	Site selection
System:	Riverine, lacustrine, palustrine
Criterion:	Locate the wetland in an area where at least part of the water source is ground water.
Rationale:	Ground water inflow into a wetland will reduce water temperatures since ground water temperatures are lower than surface water temperatures, particularly in summer months. High temperatures in the water column may be a limiting factor for aquatic organisms.
Methods:	Select a site whose water supply is at least partially from ground water. A ground water monitoring program, such as that described in Chapter 2, Wetland Hydrology, will determine if a ground water source is available for the replacement wetland in addition to its surface water source.
Notes:	This feature should be considered in conjunction with Section 9.1, Hydroperiod.

Section 9.5
Water Level Control

Type: Site selection

System: Estuarine, riverine, lacustrine, palustrine

Criterion: Avoid subjecting the wetland to drastic artificial water level fluctuations.

Rationale: Although partial drawdown can have a positive effect on aquatic community diversity, severe fluctuations will have a negative effect. Large water level fluctuations can expose spawning areas, denude vegetative cover, and generally reduce aquatic invertebrate diversity.

Wetlands located downstream from impoundments may have reduced aquatic diversity, since the impoundment may reduce the outflow of detritus, the food source for aquatic insects.

The cutoff in drawdown depth and frequency which creates a negative impact on the aquatic community is not known. However, research has determined that wetland plant communities thrive in reservoir areas where the seasonal drawdown is less than 12 feet and where adjacent shore gradients are minimal. Maximum daily water level changes of one foot do not appear to affect benthic communities (Smith et al., 1981), but fluctuations greater than 3 feet will have adverse effects (Fisher and Lavoy, 1972).

Methods: Select a site which is not within the affected upstream or downstream reach of a surface water impoundment.

Section 9.6
Channel Gradient and Water Velocity

Type: Site design

System: Riverine

Criterion: Create a gradual channel gradient with low velocity for a portion of the wetland or create a channel with velocities of less than 1.5 feet/second during the two year flood.

Rationale: Wetlands with some areas having low water velocities usually have greater fish and invertebrate diversity than wetlands with only fast flowing water. The presence of some higher velocity areas creates habitat diversity.

Methods: Keep the channel gradient low in some portions of the wetland, so as to create depositional velocity conditions in some portions of the wetland channel. Meander the channel to increase length and decrease velocity. Some velocity must be maintained to avoid water stagnation. Therefore, avoid gradients of 0.0.

See Table 2, Section 3.5, Nutrient Removal/Transformation, for channel gradients which will provide depositional velocities.

Section 9.7
Water/Vegetation Proportions

Type: Site design

System: Estuarine, riverine, lacustrine, palustrine

Criterion: Create a balance of vegetated and unvegetated areas in locations inundated by surface water.

Rationale: Some vegetation is a beneficial food source for fish. However, vegetation which is too dense reduces fish movement.

Methods: *For palustrine and estuarine systems:*
In areas where surface water is present, provide some open water and some emergent or aquatic bed vegetation. Proportions of vegetated areas to open water areas should range from approximately 70/30 percent to 40/60 percent.

Open water and emergent or aquatic bed areas can be created by varying the bottom elevation of the wetland according to the desired location of the open water and vegetated areas. Generally, unvegetated open water areas will be present at depths greater than 6.6 feet and emergent vegetation will be present in water from 0.0 to 1.0 foot in nontidal systems.

In tidal systems, emergent plants rarely exist at elevations lower than the normal low tide.

For lacustrine and riverine systems:
In the area where surface water is present, provide at least 10 percent coverage by emergent and aquatic bed vegetation.

Notes: Section 9.8, Vegetated Canopy, may not be compatible with this feature because of the extent of shading. Therefore, it may be used in place of this feature for riverine wetlands.

Section 9.8
Vegetated Canopy

Type: Site selection and site design

System: Riverine

Criterion: Create the wetland so it is partly shaded at midday.

Rationale: Moderate shading enhances aquatic diversity in riverine wetlands. Shade can greatly lower the maximum summer temperatures where water velocity is low, and submerged cover is lacking. Low overhanging vegetation will also provide cool hiding places for large fish. Finally, leaves and organic debris produced by vegetation will contribute food sources for benthic organisms.

Methods: Provide a moderate amount of shade to the wetland by partial vegetated cover on its banks (Figure 18). A major portion of the river should be shaded. Deciduous woody plantings are most desirable.

Streamside vegetation may include shrubs and small trees. Trees with mature heights in excess of 20 feet are not desirable for streamside planting.

Notes: Section 9.7, Water/Vegetation Proportions, for riverine systems, may not be compatible with this feature. Therefore, it may be used in place of this feature.

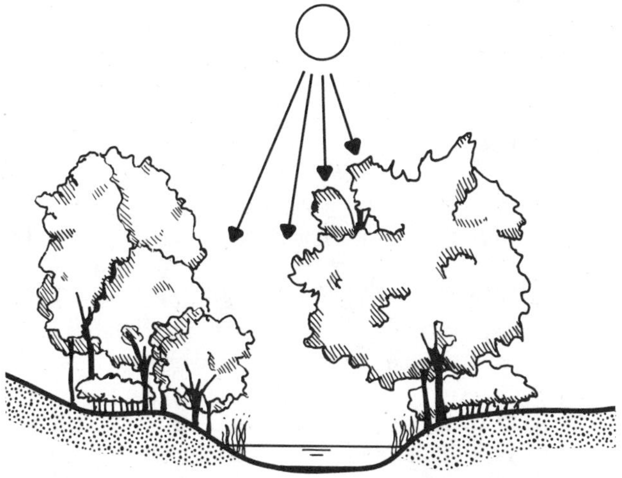

Provide a significant amount of shade using deciduous shrubs and trees on its banks. Most of the river should be shaded at midday.

Figure 18. Example of a vegetated bank for a riverine wetland.

Section 9.9
Vegetation Form Richness

Type: Site design

System: Lacustrine, palustrine

Criterion: Include several vegetation classes or subclasses interspersed with open water for the vegetated portion of the wetland.

Rationale: Plant form diversity is often associated with fish habitat diversity. It is also indicative of the number of habitat conditions and food sources available in a wetland.

Methods: In wetlands less than ten acres, include a minimum of three vegetation classes or four vegetation subclasses interspersed with open water. If the replacement wetland is greater than ten acres, an even higher number of vegetation classes and subclasses should be included. The vegetation classes should be in approximately even proportions.

Open water and vegetated areas are dependent on water depth. Vegetated and open water areas in a replacement wetland can largely be controlled by the depth of the water. Unvegetated open water which does not support either surface or submerged plants will be present at water depths of 6.5 feet and greater. Generally, the following average depth zones are correlated with plant forms, such as depicted in Figure 19:

- Trees and shrubs - 2.0 to 0 feet above normal pool elevation
- Emergent plants - 0 and 1.0 feet of water
- Rooted surface plants - 1.0 to 2.0 feet of water
- Rooted submerged plants - 1.5 to 6.5 feet of water

- Unvegetated open water is found at depths greater than 6.6 feet

Exact water depth requirements are strictly dependent on the individual plant species requirements and the duration and season of flooding. This species specific information must be known prior to determining the water depths.

The outer surface water edge of wooded wetlands supports high densities of aquatic invertebrates and fish. Broad-leaved deciduous wooded wetlands, particularly those with alder and willow species, are high detritus producers. It is therefore useful to include wooded wetlands in the wetland design, particularly adjacent to a surface water edge.

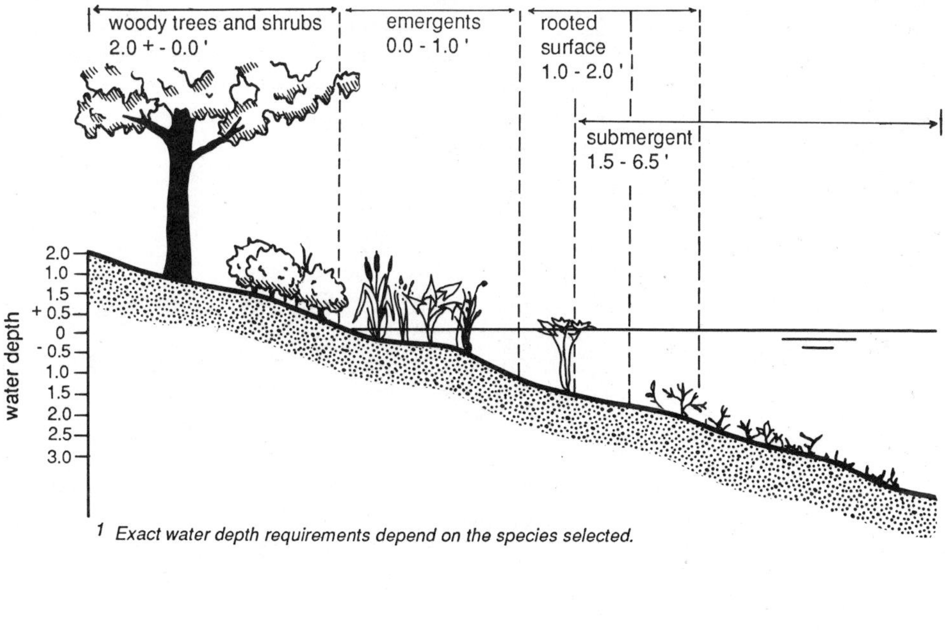

Figure 19. Generalized depth zonation of wetland plant forms for nontidal systems.

Section 9.10
Vegetation/Water Interspersion

Type: Site design

System: Estuarine

Criterion: Provide a mosaic of small patches of emergent vegetation interspersed with pools, channels, and flats.

Rationale: Wetlands containing vegetation interspersed with open water and channels have a greater diversity of fish and invertebrates. Also, the open water/wetland edge often associated with interspersion is positively correlated with higher fish populations (Zimmer and Bachman, 1978).

Methods: In areas where surface water is present at the specified tide, provide small patches of vegetation interspersed with open water by varying the bottom elevation. In estuarine systems, emergent plants are found above the elevation of the normal low tide. Below the normal low tide, a tidal system normally supports no emergent vegetation.

Some commercially available estuarine species include the following:

<u>*East Coast Emergents in Flooded Estuarine Areas*</u>

salt marsh cordgrass	*Spartina alterniflora*
saltmeadow cordgrass	*Spartina patens*
big cordgrass	*Spartina cynosuroides*
needlerush	*Juncus roemerianus*
narrow leaved cattail	*Typha angustifolia*
southern wild rice	*Zizianiopsis miliacea*
giant wild rice	*Zizania aquatica*
prairie cordgrass	*Spartina pectinata*

West Coast Emergents in Flooded Estuarine Areas

common pickleweed	*Salicornia virginica*
sea blite	*Suaeda californica*
arrow grass	*Triglochin maritima*
California cordgrass	*Spartina foliosa*

Contact with local seed sources and nurseries in the region will provide additional information regarding commercially available species. Appendix B provides a partial list of nurseries specializing in wetland plant stock.

Section 9.11
Aquatic Habitat Features

Type: Site design

System: Riverine

Criterion Include diverse habitat features such as several substrate types, water velocities, and depths.

Rationale: Some aquatic organisms prefer a specific habitat. Therefore, riverine wetlands which support a high diversity of habitat conditions will in turn support diverse aquatic fauna. Pools provide nursery areas for young fish and are used as refuges during low flow periods by adult fish. Riffle areas between pools support a high number of aquatic invertebrates.

Methods: A diversity of habitat features may be created using the following techniques:

- Create a balance of riffle and pool areas.
- Create slow water areas to include pools and backwaters. This is most easily done by creating a flow velocity generally less than 0.6 feet/second. These conditions are described in Table 2, Section 3.6, Nutrient Removal/ Transformation.
- Riffle areas between pools should contain a cobble-gravel substrate.
- Provide fish cover using moderately dense aquatic vegetation, submerged logs and stumps, boulders, and overhanging vegetation. Or, ensure that cover is available in a hydrologically connected wetland, within a distance of five times the stream width.
- Avoid straight trapezoidal channels. Incorporate meanders to encourage pool development and inside depositional point bars. Vary the channel cross section.

Some aquatic habitat improvement devices are summarized below:

Large rocks or boulders placed in the stream can be used to reduce stream velocities and to create surface turbulence and small scour holes. These rocks should be placed toward the middle of the stream so that the stream flow will not be deflected towards an unprotected bank.

Deflectors can be used to direct water current in the stream by concentrating flow and narrowing the channel. Deflectors will cause the stream to create scour holes or pools downstream and a stilling area against the streambank. Scour holes provide resting areas for fish and the upstream stilling area is highly suitable for invertebrate production. By narrowing the channel flow, deflectors have the added benefit of keeping gravel areas clean for spawning and surface turbulence provides fish hiding cover. Deflectors may be made from stones, gabions, logs, or a combination.

A half-log habitat device can provide additional fish hiding cover. A half slab of wood is placed parallel to the stream flow at midstream, and is secured by a reinforcing rod.

An elevated boulder structure is a useful means of providing fish cover and a scouring effect in large, flat stream reaches. The structure is a rectangular wood frame elevated above the stream bed by spacer logs. Rocks placed on top of the wood frame provide water turbulence and stability to the structure.

A large volume of information exists on fish habitat improvement measures for streams, and these should be consulted with regards to specific measures including the U.S.D.A. Forest Service's *Fish Habitat Improvement Handbook* (1985). Additionally, the State fish and game authority, and the regional office of the U.S. Fish and Wildlife Service should also be consulted for information on fish habitat improvement guidelines suitable to the region in which the replacement site is located.

Section 9.12
Vegetation Class

Type: Site design

System: Riverine, lacustrine, palustrine

Criterion: Include aquatic bed vegetation in at least part of the wetland.

Rationale: Densities of invertebrates are often greater in aquatic beds than in stands of emergent vegetation, probably because greater surface area is available in aquatic beds due to the dissected leaf form.

Methods: Aquatic bed vegetation (either submerged or surface) should be included in a portion of the wetland. The suggested area is no less than 10 percent of the surface area. A partial list of submerged vegetation particularly attractive to invertebrates includes the following:

curly pondweed	*Potamogeton crispus*
common hornwort	*Ceratophyllum demersum*
lesser duckweed	*Lemna minor*
star duckweed	*Lemna trisulca*

Select plants which are commercially available and native to the replacement area. A partial listing of wetland and native plant suppliers may be found in Appendix B.

Section 9.13
Physical Habitat Interspersion

Type: Site design

System: Lacustrine, palustrine

Criterion: Include a variety of substrate types, water velocities, and water depths in the wetland.

Rationale: Aquatic organisms, a food source of fish species, prefer specific habitat conditions including substrate type, water velocities, and depths. Therefore, the greater the number of habitat conditions present for aquatic organisms the greater the diversity of these organisms.

Methods: Provide as many physical habitats in the flooded portions of the wetland as possible.

Water Depth. Vary the bottom elevation. Several deep pools are desirable fish refuges for periods of low water.

Water Velocity. The gradient of the bottom elevation should be varied from the wetland's inlet to its outlet. Create slack water areas in the wetland by minimizing gradients (see Section 9.6, Channel Gradient and Water Velocity).

Substrate Types. Add a variety of substrate conditions, including gravel (particularly in higher velocity areas), large woody debris, and boulders. Sand should be omitted as a substrate type (see Section 9.16, Substrate Type).

Notes: This criteria should be combined with Section 9.9, Vegetation Form Richness, and Section 9.10, Vegetation/Water Interspersion.

Section 9.14
Diversity Enhancement

Type: Site selection and site design

System: Estuarine, riverine, lacustrine, palustrine

Criterion: Increase the habitat diversity of the site in relation to its hydrologically connected wetlands.

Rationale: A wetland replacement site should be evaluated as part of a larger wetland system. That is, the aquatic habitat qualities of the replacement site can be used to increase the diversity of the larger, hydrologically connected area. Essentially, a replacement wetland can provide missing habitat requirements thereby increasing the overall aquatic diversity value for the entire area.

Methods: Select a site which can be hydrologically connected to an existing wetland. Evaluate adjacent hydrologically connected wetland characteristics for the following:

- aquatic habitat features (Section 9.11)
- water/vegetation proportions (Section 9.7)
- vegetation form richness (Section 9.9)
- physical habitat interspersion (Section 9.13)
- hydroperiod (Section 9.1).

Methods for this evaluation are presented in each section, as noted.

Determine if the wetland replacement site, functioning as part of a larger wetland system, could add any or all of the above habitat characteristics to the existing system. Using Vegetation Form Richness (Section 9.9) as an example, if existing wetlands immediately adjacent to the replacement site support two vegetation classes (such as emergent and aquatic bed), the replacement site itself would need only two additional types of vegetation classes to provide a total of four classes for the entire wetland system. This will increase the vegetation form richness of the replacement site as well as the vegetation form richness of the existing wetlands.

Section 9.15
Watershed Size

Type: Site selection

System: Riverine, lacustrine, palustrine

Criterion: Locate the wetland replacement site adjacent to a third order or greater stream, or select a site with a watershed greater than 100 square miles.

Rationale: Diversity of aquatic communities increases as the watershed increases in size. Food is more readily available and flows are more dependable in larger order streams and low elevation wetlands, which tend to be found in larger watersheds.

Methods: Compute the size of the watershed which drains into the proposed replacement site, and determine the order of the stream on which it is located. The wetland's surface waters should be hydrologically connected to the stream (see Section 9.2, Outlet Characteristics).

If no third order stream or 100 square mile watershed is present, select the highest order stream and/or watershed possible for the replacement site.

Section 9.16
Substrate Type

Type: Site selection and site design

System: Lacustrine, palustrine

Criterion: Avoid using sand as a substrate. Also, avoid using large areas of cobble, gravel, or bedrock.

Rationale: Organic sediments generally support higher densities of fish and aquatic invertebrates. Sand generally supports low densities of macroinvertebrates, mainly because of its physical instability.

Methods: Evaluate the texture of the substrate material proposed for use. The most desirable substrates are organic materials. Consider stockpiling and using the sediments of wetlands filled by the highway project in the replacement area.

Section 9.17
Wetland Acreage

Type: Site selection

System: Lacustrine, palustrine

Criterion: Locate the wetland site so it is hydrologically connected to as many other wetland areas as possible, and that the total surface area of the replacement site and its connected wetlands exceeds 40 acres.

Rationale: Generally, the larger the surface area of the replacement site and other hydrologically connected wetlands and deep water areas, the higher the species diversity.

Methods: Select a site where the replacement wetland can be directly connected by channels to other wetlands nearby. The connections should be deep enough to allow for fish movement into and out of the replacement wetland. Generally, water depths should be greater than 4.0 inches, free of obstructions which may prevent fish passage, and have low water velocity.

Section 9.18
Sediment and Contaminant Sources

Type: Site selection

System: Riverine, lacustrine, palustrine

Criterion: Avoid locating the replacement wetland in an area where the site's receiving waters contain significant amounts of inorganic sediment.

Rationale: Aquatic diversity is greatest where the levels of suspended solids are low. Suspended solids have a negative effect on spawning areas, fish movement, invertebrate habitat, and plant productivity.

Methods: Select sites which are not likely to receive runoff from the following: stormwater outfalls, irrigation return waters, surface mines, unvegetated disturbed soils, sand or gravel pits, severely eroding stream or road banks, or soils classified by the U.S. Soil Conservation Service as having a severe erosion hazard.

Section 9.19
Water Chemistry

Type: Site selection

System: Estuarine, riverine, lacustrine, palustrine

Criterion: Ensure that the wetland site's receiving waters meet water quality standards for pH, temperature, suspended solids, dissolved solids, and dissolved oxygen.

Rationale: Aquatic productivity is directly correlated with water quality.

Methods: Conduct water quality testing for the ground water and/or surface water source for the replacement site. Water quality criteria and standards for the site should generally meet those set forth by the State in which the site is located. Principal criteria to measure include the following:

- *pH.* Circumneutral pH (between 5.6 and 8.6) is optimal for aquatic production. Acidic waters adversely affect fish populations by mobilizing toxic metals.

- *Salinity and Conductivity.* Aquatic abundance and diversity is correlated in lacustrine and palustrine wetlands with salinity less than 5 ppt and in estuarine wetlands with salinities less than 40 ppt.

- *Suspended Solids.* Low levels of suspended solids less than 80 mg/l contained in runoff and surface waters entering a wetland have a high correlation with aquatic diversity and abundance. The maximum level should not exceed 200 mg/l.

- *Dissolved Solids or Alkalinity.* A minimum of 20 mg/l $CaCO_3$ is desirable for high aquatic productivity. Moderately elevated alkalinity levels usually are associated with increased fish standing crop.

- *Dissolved Oxygen.* Dissolved oxygen concentrations greater than 4 mg/l or 60 percent saturation are optimal.

Section 9.20
Wetland/Watershed Ratio

Type: Site selection

System: Estuarine

Criterion: Locate the wetland downslope from existing wetlands in the watershed.

Rationale: A surface water connection to adjacent waters and wetlands enables access to the wetland by colonizing organisms and allows the input of natural food material.

Methods: Determine the watershed of the replacement site. Using the National Wetlands Inventory Maps, field survey, and/or aerial photographs, determine if any wetlands would drain into the replacement area. Locate the replacement wetland in a watershed which is comprised of at least 5 percent wetlands.

Notes: This feature overlaps with Section 9.14, Diversity Enhancement; Section 9.15, Watershed Size; and Section 9.17, Wetland Acreage. The requirements of these features should be included in the consideration of Section 9.20.

Chapter 10
Wetland Dependent Bird Habitat Diversity

Description of the Function:

Wetlands provide habitat for numerous species of birds, mammals, reptiles, amphibians, fish, and shellfish. Depending upon the size of the wetland, the vegetative composition, and the requirements of the specific animal, wetlands can provide all or some of a species' life requisites.

The use of wetlands by wildlife can also depend on the seasonal requirements and the life stage of the specific animal. For example, dense vegetation of wetlands can provide important winter cover for both mammals and birds. Waterfowl use specific wetland types during their specific life stages such as reproduction, molting, and migration and wintering (Mistch and Gosselink, 1986).

Because the WET evaluation for wildlife diversity and abundance is concerned principally with bird species, this portion of the guidebook is in turn limited to that group. This is not meant to suggest that wetlands are less important to other wildlife groups. Developing habitat for wetland dependent birds can and will also provide habitat needs for other wildlife groups as well.

The primary goal in providing habitat for wetland dependent birds is to create an environment which promotes diversity and abundance of these species. Although diversity and abundance are different conceptually, they are known to be highly correlated with each other.

General Design Concepts:

Like aquatic diversity, wetland dependent bird habitat diversity requires consideration of a number of site selection and site design criteria. Principal factors which require consideration include size, cover, food, specialized habitat needs, and the geometric and seasonal qualities of cover and food requirements. The most critical site selection and site design concept for this wetland function is the diversity of habitat conditions within the wetland as well as on a regional basis. A diversity of habitat conditions in turn will create a diversity of wetland dependent birds.

On a regional basis, the first step in developing a replacement design for wetland dependent birds involves an assessment of the acreage and kinds of wetlands existing in the replacement region. It also requires identifying complimentary or limiting factors present (Sections 10.10 and 10.11), particularly in drought-prone and intense storm regions. Wetland density and the proximity of a wetland to other wetlands within a region (Section 10.2) is also an important element in increasing regional wetland habitat diversity.

A variety of habitat conditions within a given wetland is also critical in creating species diversity. Interspersion of vegetation and water (Sections 10.3 and 10.4) and the length of shoreline (Section 10.18) correlate directly with bird species diversity. Structural diversity of vegetation, the number of vegetation forms, is also important (Section 10.6).

Generally, as a wetland's size increases, so does its habitat diversity and stability. However, increasing the size of small wetlands usually has a greater impact on increasing animal species richness than increasing the size of large areas (Sections 10.1 and 10.2).

Cover protects individuals from predation. Since cover needs vary from species to species, it is important to provide a variety of cover types. Breeding cover needs may be fulfilled by providing different vegetation forms (Section 10.6). However, wintering and migration needs are best met by a predominance of woody vegetation (Section 10.5). Cover availability is largely dependent on the hydroperiod.

Islands provide excellent cover (Section 10.15). Adjacent upland cover (Sections 10.7, 10.20, and 10.25) and wide stands of aquatic vegetation (Section 10.8) in a wetland can also appreciably add to cover variety and complexity. Shelter is important in protecting wintering and breeding birds from storms (Section 10.16).

Wetlands must also provide a variety of food sources to support bird diversity. The full range of feeding strategies must be provided, be they herbivorous, carnivorous, insectivorous, or omnivorous. In that no particular wetland class is better than another in providing food sources, a diversity of wetland classes should be represented (Section 10.6). For waterfowl species, rooted vascular aquatic bed vegetation is most desirable (Section 10.9).

The establishment and growth of wetland vegetation can be encouraged by providing certain physical conditions in the wetland basin. Some of these include low water velocity (Section 10.13), substrates which provide stability and nutrients for plant growth (Section 10.17), and nonacidic water which encourages plant productivity (Section 10.22).

Specific Site Selection and Site Design Features:

The manner in which wetland dependent bird species use their habitat is tied directly to specific activities which include breeding, wintering, and migration. Because specific habitat needs vary depending on the type of activity, wetland features vary accordingly. A different set of features must be emphasized for a wetland whose principal function is for breeding, in comparison to a wetland whose function is principally for migration or wintering. All three activity categories may be successfully included in many wetland replacement designs. But the individual features and their relationship to the individual activity must be evaluated in the site selection and site design processes.

Table 10 lists the features for wetland dependent bird habitat diversity. The Importance to Function category describes the relative value of each design feature to wetland dependent bird habitat diversity. In addition, it also notes the type of activity supported by each feature: breeding (B), migration (M), and wintering (W). Thus, if the principal goal is to provide breeding habitat, those features which are associated with breeding activity may be selected for use in the wetland replacement design process.

The type of activity a wetland supports also depends to a large degree on the wetland system. Riverine, lacustrine, and palustrine wetlands are most desirable for breeding activity because they are most likely to provide the diverse habitat conditions necessary for this activity. Estuarine systems generally tend to support more monotypic vegetation, and therefore provide less habitat diversity than palustrine, riverine, and lacustrine systems. Estuarine systems, as a whole, are extremely important to migration and wintering. However, an exception to this is estuarine wooded wetlands such as mangroves, which support numerous breeding birds. Because of this, the replacement wetland's system should have a bearing on the type of activity which is best emphasized in developing the design. Table 10 also lists the applicable wetland systems for each design feature.

Design features for breeding also vary according to the climatic region of the country. In dry (precipitation deficit) regions, wetlands tend to have high biotic diversity because of varying water levels, but are relatively scarce in number. The system category in Table 10 also notes the applicability of some design features to dry (precipitation deficit) areas or wet (precipitation excess) areas.

Table 10. Wetland dependent bird habitat diversity site selection and site design features.

Section	Feature	System [1]	Activity [2]	Site Selection	Site Design	Importance to Function [3]	Notes [4]
10.1	Wetland Acreage	E,R,L,P	B,M,W	X	X	Moderate	2
10.2	Wetland Density	E,R,L,P	B,M,W	X		Moderate	3 renamed
10.3	Water/ Vegetation Proportions and Interpersion	E,R,L,P	B,M,W		X	High	15.1, 31 renamed
10.4	Vegetation Class Interspersion	E,R,L,P	B,M,W		X	High	16
10.5	Vegetation Class and Subclass (Primary)	E,R,L,P	M,W		X	Low	12
10.6	Vegetation Form Richness	E,R,L,P	B,M,W		X	High	17
10.7	Vegetated Canopy	R - nontidal	B	X	X	High	20
10.8	Vegetated Width	P - dry regions	B		X	Moderate	36
10.9	Plants: Waterfowl Value	E,R,L,P	M,W		X	Moderate/High	50
10.10	Diversity Enhancement	E,R,L,P	B,M,W	X	X	High	new
10.11	Regional Diversity	E,R,L,P	B,M,W	X	X	High	38 renamed
10.12	Most Permanent Hydroperiod	P - dry regions	B	X	X	Moderate	33
10.13	Channel Gradient and Water Velocity	R - dry regions	B		X	Low	7,41 renamed
10.14	Outlet Characteristics	E,R,L,P	B,M,W		X	Low	8 renamed
10.15	Islands	E,R,L,P	B		X	Moderate	14
10.16	Fetch/Exposure	E,R,L,P	B,W	X		Moderate	19
10.17	Substrate Type	E,R,L,P	B,M,W	X	X	Low/Moderate	45
10.18	Upland/Wetland Edge	E,R,L,P	B,M,W		X	Low	18
10.19	Special Upland Habitat Features	E,R,L,P	B,M,W	X		High	39 renamed
10.20	Adjacent Forest Acreage	E,R,L,P	B	X		Moderate	2 renamed
10.21	Watershed Size	R	B	X		Moderate	4.2
10.22	pH	E,R,L,P	M,W	X		Low/Moderate	47
10.23	Contaminant Sources	E,R,L,P	B	X		High	27
10.24	Human Disturbance	E,R,L,P	B,M,W	X		High	30
10.25	Land Cover of the Watershed	E,R,L,P	B,M,W	X		Moderate	21
10.26	Location and Size	E,R,L,P	M,W	X		Moderate	4.1

1 E - estuarine, R - riverine, L - lacustrine, P - palustrine. Additionally, some features are noted as "Wet Regions" or "Dry Regions". Features listed as "Dry Regions" are applicable to areas where evaporation exceeds precipitation. "Wet Regions" apply to areas where precipitation exceeds evaporation.. However, all features should be considered regardless of location.

2 B - breeding, M - migration, W - wintering

3 These ratings are generally derived from Volume I of WET (FHWA-IP-88-029). Some values were modified in relation to wetlands replacement criteria by Paul Adamus (U.S. EPA Environmental Research Lab, Corvallis, Oregon).

4 This column describes the WET 2.0 predictors used in deriving each feature. If only a number is given, it refers to the predictor number in WET 2.0. RENAMED means the title of the WET 2.0 predictor was modified to show the new feature. NEW means the feature is not based on any given WET 2.0 predictor, but was inferred from other information.

WET 2.0 Predictors Not Used
1 Climate; 10 Wetland System; 23 Ditches/Canals/Channelization/Levees; 28 Direct Alteration; 32 Hydroperiod; 34 Water Level Control; 48 Salinity and Conductivity

Section 10.1
Wetland Acreage

Type: Site selection and site design
breeding, migration, and wintering

System: Estuarine, riverine, lacustrine, palustrine

Criterion: Size the replacement wetland so it is greater than five acres. Or, locate a replacement site that is hydrologically connected to existing nearby wetlands so that the total wetland area is greater than five acres.

Rationale: Larger wetlands, or wetlands directly connected to other wetlands, are more likely to support higher on-site diversity and abundance of wetland dependent birds than small wetlands. These wetlands generally provide a greater variety of habitat and food resources than small wetlands.

The expansion of small, existing wetlands (less than five acres in size) has a greater potential to increase species diversity than expanding larger wetland areas.

Also, expanded wetlands have a greater potential to add to the regional wetland diversity if they are located more than a mile from other large wetlands or groups of wetlands (see Section 10.11, Regional Diversity).

Methods: Create a replacement wetland whose size is a minimum of five acres. For smaller replacement wetlands, ensure that it is hydrologically connected to other nearby wetlands so that the total wetland area is a minimum of five acres.

For hydrologic connectivity between wetlands, ensure that the surface water connecting the wetlands is at least 4.0 inches in depth during the growing season, has slow water velocities, and is free of obstructions. Additionally, design the width of the connecting surface water to be a minimum of half the width of the replacement wetland or the satellite (connected) wetlands.

Section 10.2
Wetland Density

Type: Site selection
breeding, migration, and wintering

System: Estuarine, riverine, lacustrine, palustrine

Criterion: Select a site where the wetland will be the only one within a large geographic area (an oasis wetland). Or, select a site where the replacement wetland, in combination with other adjacent wetlands, is part of a dense regional cluster within the same geographic area (complex/cluster wetland). (See glossary for definitions.)

Rationale: Oasis wetlands attract wildlife from large geographic areas, and are thus significantly greater in wildlife value than wetlands scattered evenly throughout a region.

Complex/cluster wetlands are comparatively large wetlands in relation to others in the same geographic area. These wetlands attract species that avoid smaller wetlands. They also provide greater internal habitat diversity and stability than smaller wetlands.

Methods: Select a site which, in combination with other existing wetlands, meets the complex/cluster or oasis acreage criteria set forth in Table 11 and detailed below.

Oasis Situation
The focus of this design feature for oasis situations is to identify the acreage of different wetland classes in the area. An increase in the smallest class would most likely provide increased wetland dependent bird habitat diversity.

To do this, compute the acreage of existing scrub/shrub forested wetlands within 1000 feet of the replacement site in a nontidal system. Or, for a tidal system, compute the acreage

Table 11. Acreage criteria for oases and clusters for emergent (EM), and scrub/shrub forested (SS/FO) vegetation classes, and wetland loss rates .

State	Palustrine (ac/mi 2)				Estuarine (ac/mi of shoreline)				Loss Rate (% year)
	EM 1 Oasis	EM 1 Cluster	SS/FO 2 Oasis	SS/FO 2 Cluster	EM 1 Oasis	EM 1 Cluster	SS/FO 2 Oasis	SS/FO 2 Cluster	
AL	0.4	2.3	11.1	66.5	7.6	45.6	ND	ND	0.67**
AZ	0.2	1.3	1.2	7.0	---	---	---	---	0.42***
AR	0.9	5.6	9.1	54.6	---	---	---	---	1.80
CA	0.3	1.6	0.2	1.0	6.1	36.8	ND	ND	0.42***
CO	0.6	3.7	0.5	2.7	---	---	---	---	0.42***
CT	0.5	2.9	7.8	47.0	5.9	35.3	1	1	0.35**
DE	0.6	3.8	9.6	57.7	47.1	282.7	1	1	0.81
FL	11.3	67.8	21.7	129.9	27.8	166.5	13	78	0.57
GA	0.7	4.2	15.6	93.6	29.3	175.7	1	1	0.35**
ID	0.2	1.4	0.6	3.8	---	---	---	---	0.42***
IL	0.2	1.1	2.2	13.0	---	---	---	---	0.84
IN	0.4	2.6	0.8	5.0	---	---	---	---	0.67**
IA	1.3	7.6	1.6	9.7	---	---	---	---	0.67**
KS	0.3	1.9	0.2	0.9	---	---	---	---	0.42***
KY	0.2	1.1	0.4	2.3	---	---	---	---	0.67**
LA	5.3	31.8	21.4	128.6	48.8	292.9	ND	ND	0.84
ME	1.6	9.9	8.6	51.7	4.6	27.6	ND	ND	0.35**
MD	0.3	2.0	3.8	22.6	10.3	62.0	1	1	0.35**
MA	1.5	9.1	10.8	64.5	3.0	18.2	1	1	0.35**
MI	3.2	19.2	9.7	58.1	---	---	---	---	0.67**
MN	8.8	53.0	9.9	59.6	---	---	---	---	0.67**
MS	1.3	7.9	14.7	88.3	4.8	28.7	ND	ND	1.48
MO	0.2	1.4	.13	7.7	---	---	---	---	0.67**
MT	0.8	4.6	0.4	2.3	---	---	---	---	0.42***

1 EM = emergent wetlands
2 SS/FO = scrub/shrub forested wetlands

* Wetland acreage estimates were not available for this state, so data from nearby states were used. More detailed or accurate data on wetland densities from state or local agencies may be substituted if available. The following formula should be applied to improve the definition of clusters and oases: Oasis = 0.2x; Cluster = x + 0.2x (where x = mean statewide density of wetlands in acres per square mile).

** State data were statistically insignificant, and figures represent regional (flyway) data. **Substitute more detailed or accurate data if available.**

*** State data were statistically insignificant, and figures represent the national loss rate (0.42%). **Substitute more detailed or accurate data if available.**

Source: Adamus et al., 1987

Table 11 (continued). Acreage criteria for oases and clusters for emergent (EM), and scrub/shrub forested (SS/FO) vegetation classes, and wetland loss rates .

	Palustrine (ac/mi 2)				Estuarine (ac/mi of shoreline)				
State	EM 1 Oasis	EM 1 Cluster	SS/FO 2 Oasis	SS/FO 2 Cluster	EM 1 Oasis	EM 1 Cluster	SS/FO 2 Oasis	SS/FO 2 Cluster	Loss Rate (% year)
NE	3.5	21.1	1.0	5.9	---	---	---	---	0.42***
NV	0.2	1.0*	0.1	0.1*	---	---	---	---	0.42***
NH	0.6	3.6	3.0	17.8	4.3	25.6	ND	ND	0.35**
NJ	0.7	4.1	13.6	81.8	---	---	---	---	0.35**
NM	0.6	3.7	0.1*	0.1*	---	---	---	---	0.42***
NY	1.1	6.7	2.7	16.0	6.6	39.9	ND	ND	0.35**
NC	1.7	10.2	19.0	113.9	10.4	62.5	ND	ND	0.65
ND	7.1	42.7	0.5	3.1	---	---	---	---	0.42***
OH	0.7	4.4	1.2	6.9	---	---	---	---	0.67**
OK	0.4	2.6	2.5	15.1	---	---	---	---	0.42***
OR	1.6	9.7	0.8	4.6	8.7	51.9	ND	ND	0.42***
PA	0.3	1.8	1.6	9.4	---	---	---	---	0.35**
RI	0.5	3.0	7.9	47.1	16.5	99.3	ND	ND	0.35**
SC	1.3	7.8	25.1	150.8	32.4	194.3	1	1	0.35**
SD	3.2	18.9	0.2	1.1	---	---	---	---	0.42***
TN	0.4	2.3	2.9	17.4	---	---	---	---	0.67**
TX	1.1	6.4	1.0	6.1	32.9	197.6	ND	ND	0.42***
UT	0.9	5.6	0.4	2.3	---	---	---	---	0.42***
VT	0.7	4.2	4.1	24.5	---	---	---	---	0.35**
VA	0.3	1.8	3.3	19.6	14.25	85.5	ND	ND	0.35**
WA	1.6	9.7	0.8	4.6	1.8	10.7	---	---	0.42***
WV	9.2	1.0	0.5	3.2	ND	ND	ND	ND	0.35**
WI	3.2	19.2	9.9	59.3	---	---	---	---	0.67**
WY	0.7	4.2	0.4	2.3	---	---	---	---	0.42***

1 EM = emergent wetlands
2 SS/FO = scrub/shrub forested wetlands

* Wetland acreage estimates were not available for this state, so data from nearby states were used. More detailed or accurate data on wetland densities from state or local agencies may be substituted if available. The following formula should be applied to improve the definition of clusters and oases: Oasis = 0.2x; Cluster = x + 0.2x (where x = mean statewide density of wetlands in acres per square mile).

** State data were statistically insignificant, and figures represent regional (flyway) data. **Substitute more detailed or accurate data if available.**

*** State data were statistically insignificant, and figures represent the national loss rate (0.42%). **Substitute more detailed or accurate data if available.**

Source: Adamus et al., 1987

of existing scrub/shrub forested wetlands within one mile of the shoreline. Use Table 11 to determine the maximum acreage necessary for an oasis situation for scrub/shrub forested wetlands.

Given the acreage requirements of the replacement wetland, determine if its acreage, in combination with the existing scrub/shrub forested acreage, would be less than the maximum requirement for an oasis wetland situation shown on Table 11. If so, this suggests that a scrub/shrub forested vegetation class is recommended for the replacement site.

The same exercise for oasis situations may be applied to emergent wetlands.

Complex/Cluster

This design feature for complex/cluster situations also identifies the acreage of different wetland classes in the area.

Compute the acreage of existing scrub/shrub forested wetlands within 1000 feet of the replacement site. Or, for a tidal site, compute the acreage of existing scrub/shrub forested wetlands within 1 mile of the shoreline. Use Table 11 to determine the minimum acreage necessary for a complex/cluster situation for scrub/shrub forested wetlands. Given the requirements of the replacement wetland, determine if its acreage in combination with the existing scrub/shrub forested acreage would meet or exceed the minimum requirement for a complex/cluster wetland situation. If so, this suggests that a scrub/shrub forested vegetation class would be the recommended vegetation type for the replacement site.

The same exercise for complex/cluster situations may be applied to emergent wetlands.

Section 10.3
Water/Vegetation Proportions and Interspersion

Type: Site design
breeding, migration, and wintering

System: Estuarine, riverine, lacustrine, palustrine

Criterion: Create a well interspersed balance of open water and vegetation.

Rationale: The highest bird numbers and bird diversities are associated with a relatively even balance of open water which is interspersed with emergent vegetation. Open water and vegetation contact zones provide edge habitat, protection, cover, food, and territorial boundaries. Interspersion increases the vegetation/water edge zone. These contact zones between water and vegetation provide cover for breeding waterfowl. This interface also increases the amount of habitat available to species requiring vegetation and those requiring open water, which in turn increases diversity. Also, some species are specifically adapted to this edge zone.

Methods: Create a wetland that provides equal proportions of water and vegetation. Water and vegetation should be well interspersed to provide the maximum edge length or shoreline. See Figure 16, Section 8.2, Production Export.

Notes: If this feature is used in precipitation deficit regions, Section 10.6, Vegetation Form Richness and Section 10.4, Vegetation Class Interspersion, need not be considered.

If this feature is used for lacustrine and palustrine systems in precipitation excess areas, Section 10.4, Vegetation Class Interspersion need not be considered.

Section 10.4
Vegetation Class Interspersion

Type: Site design
breeding, migration, and wintering

System: Estuarine, riverine, lacustrine, palustrine

Criterion: Create a highly interspersed mosaic of relatively small areas of different vegetation classes.

Rationale: Several cover types are often required for food, shelter, nesting, lodging, and predator protection by many species. Interspersion also increases the amount of edge habitat, an important element in habitat diversity.

Methods: Provide varied bottom contours within the wetland. Create integrated patches of different vegetation classes as shown in Figure 20. This can be accomplished by planting or spreading sediments containing seeds.

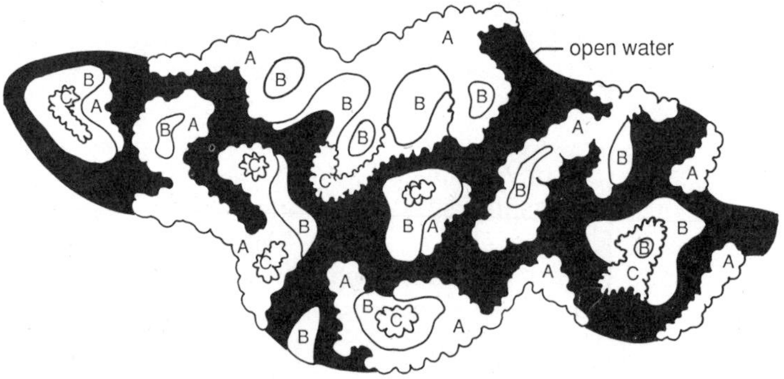

Figure 20. Example of a highly interspersed mosaic of vegetation classes.

Areas should generally exceed 100 square feet in size but not be larger than one acre.

Notes:

If this feature is used in precipitation deficit regions, Section 10.3, Water/Vegetation Proportions and Interspersion, and Section 10.6, Vegetation Form Richness, need not be considered.

If this feature is used for lacustrine and palustrine systems in precipitation excess areas, Section 10.3, Water/Vegetation Proportions and Interspersion, need not be considered.

Section 10.5
Vegetation Class/Subclass (Primary)

Type: Site design
migration and wintering

System: Estuarine, riverine, lacustrine, palustrine

Criterion: Include a vegetative cover that is predominantly forested and/or scrub/shrub.

Rationale: Vertical layering and horizontal overlap of habitats of forested or scrub/shrub wetlands support a greater diversity of wildlife species than a wetland with a less complex vegetative structure.

Methods: Include tree and shrub wetland species with relatively high wildlife values for winter food and cover. This can be determined by conducting a literature search for native wetland trees and shrubs which provide high values for winter food and cover. A partial list of vegetation species having high wildlife value is included in the following:

Wetland Trees
Deciduous

swamp white oak	*Quercus bicolor*
willow oak	*Quercus phellos*
laurel oak	*Quercus laurifolia*
pin oak	*Quercus palustris*
hackberry	*Celtis laevigata*
bitter pecan	*Carya aquatica*
parsley haw	*Crataegus marshallii*

Needle-leaved Evergreen

Atlantic white cedar	*Chamaecyparis thyoides*
black spruce	*Picea mariana*
Northern red cedar	*Thuja occidentalis*

Broad-leaved Evergreen

red mangrove	*Phyzophora mangle*
black mangrove	*Avicennia germinans*
white mangrove	*Laguncularia racemosa*
buttonwood	*Conocarpus erecta*

Wetland Shrubs

Deciduous

arrowwood	*Viburnum dentatum*
winterberry	*Ilex verticillata*
speckled alder	*Alnus rugosa*
common alder	*Alnus serrulata*
hawthorne	*Crateagus spp.*
blueberry	*Vaccinium spp.*

Broad-leaved Evergreen

rosebay rhododendron	*Rhododendron maximum*
sheep laurel	*Kalmia angustifolia*
leatherleaf	*Chamaedaphne calyculata*
cranberry	*Vaccinium oxycoccus*
Labrador tea	*Ledum groenlandicum*
wax myrtle	*Myrica cerifera*

Other tree or shrub species may be used depending on the geographic location of the site. In most cases, native, nursery grown stock or seed is desirable. Contact local districts of the U.S.D.A. Soil Conservation Service and U.S. Fish and Wildlife Service, as well as arboretums and nurseries to determine if the selected species are commercially available and native to the region.

A partial listing of wetland and native plant suppliers may be found in Appendix B.

Notes: This feature should be considered in conjunction with Section 10.6, Vegetation Form Richness.

Section 10.6
Vegetation Form Richness

Type: Site design
breeding, migration, and wintering

System: Estuarine, riverine, lacustrine, palustrine

Criterion: Include several classes or subclasses of vegetation in the wetland.

Rationale: The diversity of birds occupying a wetland is directly related to the number of vertical layers in the wetland. Complex foliage height diversity increases the number of niches available for bird breeding, feeding, and cover.

Methods: For methods, see Section 9.10, Vegetation/Water Interspersion.

Shrub and tree forms intermixed with other plant forms have a particularly significant effect on breeding bird diversity.

Additionally, aquatic bed vegetation is highly desirable as a waterfowl food source. Therefore, the shrub, tree, and aquatic bed classes are desirable to include in the replacement area.

Notes: If this feature is used in precipitation deficit regions, Section 10.3, Water/Vegetation Proportions and Interspersion, and Section 10.4, Vegetation Class Interspersion, need not be considered.

If this feature is used in lacustrine and palustrine systems in precipitation excess areas, Section 10.3, Water/Vegetation Proportions and Interspersion, need not be considered.

This feature should be used in conjunction with Section 10.5, Vegetation Class/ Subclass (Primary).

Section 10.7
Vegetated Canopy

Type: Site selection and site design
breeding

System: Nontidal riverine

Criterion: Locate the replacement site adjacent to an upland buffer consisting of trees and shrubs. Or, provide a wide upland buffer along the streambank.

Rationale: Wide vegetation buffers provide breeding cover for waterfowl and provide travel corridors for many wildlife species. In addition, vegetation buffers moderate wetland microclimate and reduce sediment laden runoff from reaching the stream.

Methods: Locate the replacement site adjacent to, or plant a vegetated buffer along the shoreline. The buffer should be a minimum of 200 feet wide, but 600 to 1000 feet is more desirable. The buffer should consist principally of trees and shrubs.

Section 10.8
Vegetated Width

Type: Site design
breeding

System: Palustrine - dry regions

Criterion: Include stands of woody and/or emergent vegetation in the wetland.

Rationale: Vegetation provides breeding cover for wildlife. It also buffers the wetland's microclimate and reduces nonpoint pollution.

Methods: In areas where vegetation and open water are interspersed, include stands of vegetation which are a minimum of 20 feet wide. Where no open water is present, make the average width of vegetation perpendicular to the shoreline exceed 20 feet.

Section 10.9
Plants: Waterfowl Value

Type: Site design
migration and wintering

System: Estuarine, riverine, lacustrine, palustrine

Criterion: Ensure that a portion of the wetland includes vegetation preferred by waterfowl.

Rationale: Wetlands providing food and cover attractive to over-wintering wetland waterfowl are likely to attract other species.

Methods: Include vegetation preferred by waterfowl in at least 10 percent of the wetland, or one acre. This vegetation must include aquatic bed or emergent species.

A partial list of species preferred by waterfowl include the following:

Aquatic Bed Species

watershield	*Brasenia schreberi*
coontail	*Ceratophyllum demersum*
little duckweeds	*Lemna spp.*
naiads	*Najas spp.*
spatterdocks	*Nuphar spp.*
water lilies	*Nymphaea spp.*
smartweeds	*Polygonum spp.*
pondweeds	*Potamogeton spp.*
water cress	*Rorippa spp.*
wigeongrass	*Rupia maritima*
big duckweeds	*Spirodela spp.*
wild celery	*Vallisneria spp.*
watermeals	*Wolffia spp.*
eelgrass	*Zostera marina*

Emergent Species

water hemp	*Acnida cannabinus*
sedges	*Carex spp.*
seashore saltgrass	*Distichlis spicata*
spikerushes	*Eleocharis spp.*
horsetails	*Equisetum spp.*
rushes	*Juncus spp.*
rice cutgrass	*Leersia oryzoides*
hooded arrowhead	*Lophotocarpus calycinus*
panic grasses	*Panicum spp.*
bull paspalum	*Paspalum boscianum*
arrow arum	*Peltandra virginica*
woody glasswort	*Salicornia virginica*
bulrushes	*Scirpus spp.*
bristle grasses	*Setaria spp.*
burreeds	*Sparganium spp.*
cordgrasses	*Spartina spp.*
annual wildrice	*Zizania aquatica*

The above listing is for the northeast United States. Other species may be applicable depending on geographic location. Consult the appropriate State wildlife agency.

Choose plant species which are native to the region and are commercially available. Contact local environmental agencies and organizations, arboretums, and nurseries for information specific to the wetland replacement site for the particular region. A partial listing of wetland and native plant suppliers may be found in Appendix B.

Section 10.10
Diversity Enhancement

Type: Site selection and site design
breeding, migration, and wintering

System: Estuarine, riverine, lacustrine, palustrine

Criterion: Increase habitat diversity of the site in relation to other hydrologically connected wetlands in the region.

Rationale: A wetland replacement site should be evaluated as part of the regional wetland resource. That is, the site's wildlife habitat qualities can be used to increase regional wetland diversity. If a replacement wetland provides a missing habitat element, it will increase the overall wildlife diversity value for the entire area.

Methods: Evaluate regional wetlands, especially nearby wetlands, to determine which characteristics are scarce. Items to consider include:

- Water/Vegetation Proportions and Interspersion (Section 10.3)
- Vegetation Class Interspersion (Section 10.4)
- Vegetation Form Richness (Section 10.6)
- Most Permanent Hydroperiod (Section 10.12).

Methods for the evaluation are presented in each section, as noted.

Determine if the wetland replacement site, functioning as part of the larger wetland system, could provide the limiting factor to any or all of the above habitat characteristics to the existing system. Using Section 10.6, Vegetation Form Richness, as an example, if existing wetlands immediately adjacent to the replacement site support two vegetation classes (such as emergent and aquatic bed), the replacement site itself would need only two additional vegetation classes to provide a total of four classes for the entire wetland system to reach a desirable level of vegetation form richness. This will increase the overall vegetation form richness of the replacement site as well as the existing wetlands.

Section 10.11 Regional Diversity

Type: Site selection and site design
breeding, migration, and wintering

System: Estuarine, riverine, lacustrine, palustrine

Criterion: *Precipitation Excess Region:*
Select a wetland system which is different than wetlands in the same region.

Arid Region:
Select an area which supports different hydroperiods than the replacement wetland. Or, select a hydroperiod for the replacement wetland which is different than wetlands in the same region.

Rationale: Regional wetland diversity is essential to maintaining wildlife diversity. Combinations of several wetland types in close proximity to the replacement site enable wildlife to expend less energy in meeting basic life requisites.

Particularly in dry regions, a diversity of wetlands with different hydroperiods (particularly more permanent hydroperiods) is important.

Methods: Select a replacement site which will provide one of the following:

<u>Precipitation Excess Regions</u>

Breeding and Wintering:
If the replacement site is estuarine, locate it within a five mile radius of a freshwater palustrine wetland, a freshwater lacustrine wetland, or a coastal island. If the replacement wetland is freshwater palustrine or lacustrine, locate the site within a five mile radius of estuarine or marine wetlands.

Wintering and Migration:
If the replacement wetland contains emergent vegetation, locate it adjacent to a mudflat.

Arid Regions

Breeding - Lacustrine and Palustrine Wetlands:
Select a site which is located within one mile of existing wetlands with different hydroperiods. Combinations of replacement wetland and regional wetland hydroperiods include the following:

- seasonally flooded,
- semipermanently flooded, and
- permanently flooded, intermittently exposed, and artificially managed for wildlife.

The wetland hydroperiods described above should cover a minimum of one acre of wetland.

Breeding - Southwestern Riparian Wetlands:
If the replacement wetland will support a principally cottonwood-willow community, it should be located immediately adjacent to a honey mesquite community. Or, the reverse situation should be true. Both wetland vegetation types should be a minimum of one acre.

Notes: This feature is most important to precipitation deficit regions and tidal systems. For precipitation excess regions, regional diversity of vegetation types (Section 10.6, Vegetation Form Richness) and wetland size (Section 10.1, Wetland Acreage) are more important.

Section 10.12
Most Permanent Hydroperiod

Type: Site selection and site design
breeding

System: Palustrine - dry regions

Criterion: Include a semipermanently flooded, seasonally flooded, or irregularly exposed nontidal hydroperiod as the most permanent hydrologic regime within a basin.

Rationale: Wetlands where surface water is present at some time, but are not permanently flooded, are most desirable in arid climates. Many emergent plants germinate only in shallow water or mud flats, so periodic absence of water is necessary to maintain cover.

Methods: See Chapter 2, Wetland Hydrology, for determining semipermanently flooded, seasonally flooded, and irregularly exposed hydroperiods. At least one acre of the wetland should include one of these hydroperiods.

If the replacement area is located within one mile of a wetland which includes one of these three hydroperiods, the replacement area's most permanent hydroperiod should be permanently flooded or intermittently exposed nontidal.

Section 10.13
Channel Gradient and Water Velocity

Type:	Site design breeding
System:	Riverine - dry regions
Criterion:	Create a gradual wetland gradient with low water velocity.
Rationale:	Slow water velocities enable sediments carried in suspension to settle. The deposition of sediment aids in the establishment of wetland vegetation. Vegetation, in turn, provides the source of food and cover for many wildlife species.
Methods:	See Section 3.5, Nutrient Removal/Transformation, for methods.

Section 10.14
Outlet Characteristics

Type: Site design
breeding, migration, and wintering

System: Estuarine, riverine, lacustrine, palustrine

Criterion: Create a permanent surface water outlet for the wetland.

Rationale: Toxins tend to accumulate in wetlands without permanent outlets. Concentrations of toxins may limit aquatic productivity, resulting in a loss of wildlife food and cover.

Methods: A permanent outlet should be used so that some surface water flow can be maintained throughout the year.

See Section 3.6, Nutrient Removal/Transformation, for methods.

Section 10.15
Islands

Type:	Site design breeding
System:	Estuarine, riverine, lacustrine, palustrine
Criterion:	Include one or more islands inside the wetland boundaries.
Rationale:	Islands are excellent refuges from predation, and have a high degree of vegetation/water edge zone and upland/wetland edge (Section 10.3 and Section 10.18, respectively). Islands also provide close proximity to all essential breeding requirements of water, food, and cover.
Methods:	Place one or more islands inside the wetland boundaries. The islands should be greater than 25 square feet in size, more than 50 feet from the wetland's shoreline, and separated by permanent deep water. If the wetland includes more than one island, they should be more than 50 feet apart. Bank slopes of the islands should be very gradual.

10.16 Fetch/Exposure

Type: Site design
breeding and wintering

System: *For breeding:*
Estuarine

For wintering:
Estuarine, riverine, lacustrine, palustrine

Criterion: Locate the wetland in a sheltered area.

Rationale: Sheltered areas, being free of wave scour, provide a greater abundance of food and cover. Sheltered areas are also important in winter, protecting animals from wind and precipitation.

Methods: For methods, see Section 4.4, Sediment/Toxicant Retention.

Section 10.17
Substrate Type

Type: Site selection and site design
breeding, migration, and wintering

System: Estuarine, riverine, lacustrine, palustrine

Criterion: Include a substrate consisting primarily of organic soils.

Rationale: Organic soils contain the nutrients and structure necessary to support aquatic productivity which in turn provides food and cover for wildlife.

Methods: Select a site which has organic soil. Stockpile topsoil from the wetland replacement site for use as substrate in the created wetland. Or, consider stockpiling and reusing the substrate of wetlands taken by the highway project in the replacement area.

Section 10.18
Upland/Wetland Edge

Type: Site design
breeding, migration, and wintering

System: *For breeding:*
Lacustrine, palustrine - wet regions

For migration and wintering:
Estuarine, riverine, lacustrine, palustrine

Criterion: Create an irregular edge between the wetland and upland.

Rationale: The more irregularity in the wetland/upland edge, the more edge habitat (ecotone) present. Edge habitat provides habitat for wetland species as well as those species inhabiting adjacent upland regions. In addition, these zones are also inhabited by species adapted specifically to the edge environment. Edge habitat, therefore, has an important influence on wildlife diversity.

Methods: Create a sinuous wetland/upland edge. This type of boundary may be created by grading the wetland boundary so that small upland areas weave in and out of the wetland (Figure 21).

Islands can be used to increase the upland/wetland edge. Methods for creating islands may be found in Section 10.15, Islands.

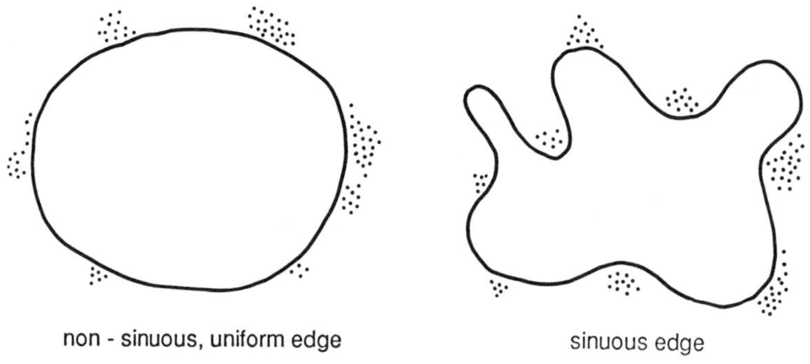

Figure 21. Sinuous upland/wetland edge.

Section 10.19
Special Upland Habitat Features

Type: Site selection
breeding, migration, and wintering

System: *For breeding:*
Lacustrine, palustrine - wet regions

For migration and wintering:
Estuarine, riverine, lacustrine, palustrine

Criterion: Place the wetland in a site containing fruit, cone, or mast bearing trees; large diameter trees; trees with snags and cavities; tilled land with waste grains; evergreens; and native prairies in the replacement area.

Rationale: All of the above are habitat features important to wildlife breeding, food, or cover needs.

Methods: Select a replacement site which has one or more of the following features within 300 feet:

- trees with snags and cavities larger than 2.0 inches (breeding),
- trees with a diameter greater than 10.0 inches (nesting),
- fruit bearing trees or shrubs (food),
- mast bearing trees or shrubs (food),
- cone bearing trees or shrubs (food),
- evergreens with over 80 percent canopy closure (cover),
- native prairie (cover and food), and
- tilled land with waste grain (food).

These items can also be included in the replacement site design or in adjacent upland areas.

Section 10.20
Adjacent Forest Acreage

Type: Site selection
breeding

System: Estuarine, riverine, lacustrine, palustrine

Criterion: Locate the replacement site adjacent to a forested area in excess of five acres.

Rationale: Species richness and density of songbirds increase with the size of the adjacent forested tract.

Methods: Locate the replacement site adjacent to a forested area which is larger than five acres in size. The forested area may be located up to one mile from the wetland site as long as it is connected by an unbroken forested corridor having a minimum width of 300 feet.

It is essential that the corridor and the forested tract are permanent (e.g., that they are located on land whose use and land cover will not be modified in the future). Examples of a permanent forest tract would include public or private nature preserves and national parklands.

Section 10.21
Watershed Size

Type: Site selection
breeding

System: Riverine

Criterion: Locate the wetland in a watershed greater than one square mile in size.

Rationale: Wetlands having large watersheds are likely to be sustained hydrologically throughout time, and are also higher in nutrients than those in smaller watersheds.

Methods: Select a site in a watershed greater than one square mile in size.

Section 10.22
pH

Type:	Site selection migration and wintering
System:	Estuarine, riverine, lacustrine, palustrine
Criterion:	Avoid acidic (low pH) water in the wetland.
Rationale:	The buffering ability of nonacidic water is better than acidic water, which generally results in higher productivity.
Methods:	Select a site where the surface and/or ground water has a pH greater than 6.0. Water testing will be necessary to determine pH.

Section 10.23
Contaminant Sources

Type: Site selection
breeding

System: Estuarine, riverine, lacustrine, palustrine

Criterion: Select a replacement site which is free of waterborne contaminants.

Rationale: A contaminated water source will reduce wildlife populations by causing mortality and decreased productivity. Toxins may indirectly affect habitat by decreasing food sources.

Methods: Both the surface and ground water source for the wetland should be free of hazardous substances. This is best determined directly by water quality testing and monitoring.

Also, direct observations of the following should be made:

- industrial and sewage outfalls,
- mines,
- landfills,
- pesticide treated areas,
- oil runoff,
- irrigation return water, and
- heavily traveled highways.

Select a location which will not receive runoff from the above listed contaminant sources.

Section 10.24
Human Disturbance

Type:	Site selection breeding, migration, and wintering
System:	Estuarine, riverine, lacustrine, palustrine
Criterion:	Minimize human disturbance in or near the wetland.
Rationale:	Human disturbance in wetland areas interferes with wildlife use. Breeding can be disturbed by people traveling on foot or by boat. Food consumption demands increase when human activities disturb nesting or resting waterfowl.
Methods:	Select a replacement site which is relatively free of disturbance by foot, boat, or vehicles of any type. Although the acceptable distance to minimize disturbance to wildlife depends on the species, a minimum of 600 feet is a recommended guideline for the distance between the wetland and significant human activity.

Section 10.25
Land Cover of the Watershed

Type: Site selection
breeding, migration, and wintering

System: Estuarine, riverine, lacustrine, palustrine

Criterion: Avoid watersheds which are primarily urban or suburban. Locate the replacement wetland in a watershed which is predominantly agricultural.

Rationale: Natural cover, including agricultural areas, provides essential cover and feeding areas outside the wetland area for migratory and wintering wildlife. Urban and suburban land uses discourage wildlife use, although wetlands placed in these settings may attract wildlife appreciated by a larger number of people.

Cultivated areas provide additional food for migratory and wintering waterfowl.

Methods: Evaluate aerial photographs of the watershed in which the potential replacement site is located. Determine the percentage cover of impervious areas (urban and suburban areas), agriculture, forest, and scrub land.

Section 10.26
Location and Size

Type: Site selection
migration and wintering

System: Estuarine, riverine, lacustrine, palustrine

Criterion: If feasible, locate the replacement site in close proximity to a major river, the Great Lakes or tidal waters.

Rationale: Migratory waterfowl use lowland river courses and tidal wetlands. Tidal wetlands are important wintering areas, and the Great Lakes are important migration corridors.

Methods: If possible, select a replacement site within close proximity to one of the following:

- a river over 100 miles in length,
- the Great Lakes, or
- tidally influenced waters.

Chapter 11
Designing for Multiple Functions

In general, wetland mitigation plans will be developed for the purpose of replacing several wetland functions. If a wetland is to be replaced on a functional basis, it is recommended that those functions which were determined to have the highest value in the affected wetland form the goals of the replacement plan. In most cases, more than one of the functions evaluated by WET will be selected.

In addition, it may often be desirable to create a wetland with one or more goals which are unrelated to the affected wetland. When functional goals are selected, some consideration should be given to the compatibility of these goals. Table 12 demonstrates the level of compatibility between the eight functions described in this guidebook. For example, Production Export may not be compatible with the Nutrient Removal/Transformation function. This table can be used as a preliminary screening device in making decisions for the selection of functional goals. At a more detailed level, the criteria for features applicable to each of the two functions may conflict. This difference in criteria for site selection and site design is also an indication that the two functions may not be compatible.

Once the functional goals of the replacement wetland have been determined, the selection of a replacement site may be considered. By applying the criteria described for site selection features, many sites can be evaluated and compared. A site selection screening process may be completed by an individual or team in order to arrive at the selection of an appropriate site.

11.1 Site Selection

Table 13 summarizes the site selection features included in this guidebook by function. For example, eight site selection features are listed for the Sediment/Toxicant Retention function. The criteria for these features, which includes land cover of the watershed and sediment and contaminant sources as

Table 12. Compatibility of functions.

	Ground Water Recharge	Floodflow Alteration	Shoreline Stabilization	Sediment/ Toxicant Retention	Nutrient Removal/ Transformation	Production Export	Aquatic Diversity/ Abundance	Wetland Dependent Bird Habitat Diversity
Ground Water Recharge	0	+	0	≠	0	0	0	0
Floodflow Alteration	+	0	+	0	+	0	0	0
Shoreline Stabilization	0	+	0	+	+	≠	≠	≠
Sediment/ Toxicant Retention	0	+	+	0	+	0	≠	≠
Nutrient Removal/ Transformation	+	+	+	+	0	≠	0	0
Production Export	≠	0	0	≠	0	0	+	0
Aquatic Diversity/ Abundance	≠	+	0	0	0	0	0	+
Wetland Dependent Bird Habitat Diversity	≠	+	+	0	0	0	0	0

Notes: Compatibility of column(vertical) functions with row(horizontal) functions in the same wetland: + - compatibility, ≠- probable conflict, 0 - no significant interaction or effect unknown. These interactions are not necessarily symmetrical. For example, nutrient removal/transformation makes a wetland less effective for production export, whereas performing for production export does not necessarily make the wetland less effective for nutrient removal/transformation. Compatibility between sediment/toxicant retention and production export has been changed from "no significant interaction" to "probable conflict". Compatibility between production export and sediment/toxicant retention has been changed from "no significant interaction" to "probable conflict".

Source: Modified from Adamus et al., 1987

TABLE 13. Summary of site selection feature importance to function.

Feature	Nutrient Removal/ Transformation	Sediment/ Toxicant Retention	Shoreline Stabilization	Floodflow Alteration	Ground Water Recharge	Production Export	Aquatic Diversity	Wetland Dependent Bird Habitat Diversity [1]
Wetland System		M (E,L,P)		M (R,L,P)	H (R,L,P)			
Fetch/Exposure		M (E,L,P)	M (E,R,L,P)			M (E,R,L,P)		M (E,R,L,P)
Water Source	H (E,R,L,P)	H (E,L,P)			H (R,L,P)		M (R,L,P)	
Flooding Extent/ Duration		M (L,P)				M/H (E,R,L,P)	H (R,L,P)	
Hydroperiod							H (E,R,L,P)	
Water Level Control						M (E,R,L,P)	M/H (E,R,L,P)	
Watershed Size						M (E,R,L,P)	L (R,L,P)	M (R)
Regional Diversity								H (E,R,L,P)
Vegetated Canopy							H (R)	H (R)
Wetland/ Watershed Ratio	M (E,R,L,P)	H (E,L,P)		H (R,L,P)		L (E,R,L,P)	L (E)	
Land Cover of Watershed		M (E,L,P)		M (R,L,P)	M (R,L,P)			M (E,R,L,P)
Sediment and Contaminant Source		M (E,L,P)					L (R,L,P)	H (E,R,L,P)
Substrate Type		H (E,L,P)				M/H (R,L,P)	M (R,L,P)	L/M (E,R,L,P)
Wetland Acreage							M (L,P)	M (E,R,L,P)
Wetland Density								M (E,R,L,P)
Location and Size								M (E,R,L,P)
Watershed Soils				M (R,L,P)	H (R,L,P)			
Underlying Strata					M (R,L,P)			
Underlying Soils					H (R,L,P)			
pH						M (E,R,L,P)		L/M (E,R,L,P)
Soils and Water Alkalinity	H (E,R,L,P)							
Diversity Enhancement							H (E,R,L,P)	H (E,R,L,P)
Special Upland Habitat Features								H (E,R,L,P)
Adjacent Forest Acreage								M (E,R,L,P)
Human Disturbance								H (E,R,L,P)
Local Topography					M (R,L,P)			
Water Chemistry					M (R,L,P)		M (E,R,L,P)	
Erosive Conditions			M (E,R,L,P)					
Resource Protection			M (E,R,L,P)					
Target Area						H (E,R,L,P)		
Nutrient Sources	M (E,R,L,P)							
Shoreline Geometry			H (E,R,L,P)					

NOTES: Importance to function: L - Low; M - Moderate; H - High; Wetland System: E - Estuarine; R - Riverine; L - Lacustrine; P - Palustrine

1 See Chapter 10, Table 10 to determine if feature applies to wet or dry regions or to breeding, migration or wintering activities.

well as others, should be applied to each potential site to determine the site's ability to provide this function.

Multiple functions may also be considered during the site selection process. In this situation, the site selection features for each of the functions may be applied to each potential site. For example, given the functional goal of replacement for both Sediment/Toxicant Retention and Floodflow Alteration, a total of 12 site selection features would be considered. As is indicated in Table 13, the wetland system feature is common to both functions, although the actual criteria for each may be different. In selecting a site, it may be desirable to apply both wetland system features to each site in order to determine if the site meets one or both of the criteria.

For a predetermined site, it may be desirable to review all of the 32 site features and their criteria to identify the specific functions which may be re-created on the site. Given one site, the opportunity to provide value for some functions will be greater than for others. The screening process at this point would identify the goals of the wetland creation project.

In addition to the site selection features described here, many other factors may guide the site selection process. At the present time, many Federal and State agencies have established policies or guidelines for wetland replacement. Some of these policies, such as replacement on an acre for acre basis, consideration of replacement sites adjacent to the area of impact, and the selection of a site within the same watershed as the affected wetland, must be considered in the site selection process. It may be advisable to employ a team approach in the determination of functional goals and during the site selection, or to solicit comments from the applicable State and Federal regulatory agencies before proceeding to the design phase of a wetland mitigation plan.

11.2 Site Design

Having selected a site which provides a level of opportunity for one or more functions, the site design features may then be considered. Table 14 summarizes the site design features according to function.

TABLE 14. Summary of site design feature importance to function.

Feature	Nutrient Removal/ Transformation	Sediment/ Toxicant Retention	Shoreline Stabilization	Floodflow Alteration	Ground Water Recharge	Production Export	Aquatic Diversity	Wetland Dependent Bird Habitat Diversity [1]
Channel Gradient and Water Velocity	M (E,R,L,P,)	H (E,L,P)				M/H (E,R,L,P)	M (R)	L (R)
Outlet Characteristics	H (E,R,L,P)	H (E,L,P)		H (R,L,P)	M (R,L,P)	H (E,R,L,P)	M (L,P)	L (E,R,L,P)
Water Depth		M (E,L,P)						
Water Source	H (E,R,L,P)	H (E,L,P)			H (R,L,P)			
Flooding Extent/ Duration		M (L,P)				M/H (R,L,P)	H (R,L,P)	
Hydroperiod	M (E,R,L,P)						H (E,R,L,P)	
Sheet Flow			M (E,R,L,P)	M (R,L,P)		H (E,R,L,P)		
Regional Diversity								H (E,R,L,P)
Artificial Drainage Features					H (R,L,P)			
Vegetated Width	M (E,R,L,P)	H (E,L,P)	H (E,R,L,P)			M (R)		M (P)
Vegetation Class		H (E,L,P)	M (E,R,L,P)	M (R,L,P)			H (R,L,P)	L (E,R,L,P)
Water/ Vegetation Proportions and Interspersion		M (E,L,P)		M/H (R,L,P)		M (E,L,P)	H (E)	H (E,R,L,P)
Vegetation Class and Form Richness	M (E,R,L,P)					M (E,R,L,P)		
High Plant Productivity						M/H (E,R,L,P)		
Vegetation Class Interspersion								H (E,R,L,P)
Vegetated Canopy							H (R)	H (R)
Plants: Waterfowl Value								M (E,R,L,P)
Water/ Vegetation Proportions			H (E,R,L,P)	M (R,L,P)			H (E,R,L,P)	
Vegetation Form Richness							H (L,P)	H (E,R,L,P)
Substrate Type		H (E,L,P)				M/H (R,L,P)	M (L,P)	L/M (E,R,L,P)
Underlying Soils					H (R,L,P)			
Wetland Acreage								M (E,R,L,P)
Soils and Water Alkalinity	H (E,R,L,P)							
Diversity Enhancement							H (E,R,L,P)	H (E,R,L,P)
Fringe/Islands Wetland						M (P)		M (E,R,L,P)
Upland/ Wetland Edge								L (E,R,L,P)
Aquatic Habitat Features							M (R)	
Physical Habitat Interspersion							M (L,P)	
Wetland/ Watershed Ratio		H (E,L,P)						
Shoreline Geometry			H (E,R,L,P)					

NOTES: Importance to function: L - Low; M - Moderate; H - High; Wetland System: E - Estuarine; R - Riverine; L - Lacustrine; P - Palustrine

1 See Chapter 10, Table 10 to determine if feature applies to wet or dry regions or to breeding, migration, or wintering actvities.

Again, the features which apply to each of the functions desired should be evaluated. Using the example previously described for site selection, 13 site design features would be evaluated for the two functions. Two of the features, outlet characteristics and wetland class, are common to both the Sediment/Toxicant Retention and Floodflow Alteration functions. The criteria for outlet characteristics for each function as described in Sections 4.3 and 6.3 are compatible. However, the criteria for vegetation class for these functions as described in Sections 4.7 and 6.8 may not necessarily be compatible.

When it is not possible to include all site design features within one wetland, some of the features must be selected at the expense of others. The applicability of a specific feature to the overall design should be evaluated. In addition, the feature's importance to the function should also be considered. Tables 13 and 14 indicate the importance to function values for each feature. In the example of the vegetation class feature, Table 14 indicates that this feature is of high importance to Sediment/ Toxicant Retention and is of moderate importance to Floodflow Alteration. This information may be used to weigh the decision between these two conflicting features.

Ultimately, a variety of site design features will be blended and incorporated into the wetland mitigation plan in order to maximize the wetland's potential for the desired functions. Although this information is intended to assist in optimizing wetland functional values, it does not ensure that a high or moderate value will necessarily be realized for any function. The information for each feature enables the designer or team to select and design for high value on the basis of the WET predictors.

11.3 Example of a Multiple Function Design

A WET evaluation is completed for an existing wetland to be filled during construction of a highway project. The wetland to be filled is determined to have high values for the following functions:

- Nutrient Removal/Transformation,
- Wildlife Diversity/Abundance (or Wetland Dependent Bird Habitat Diversity), and

- Floodflow Alteration.

These three functions are present in the existing wetland, therefore compatibility of the three functions is not a consideration.

As a result of existing agency guidelines for the region, the wetland replacement site selection and design must incorporate the following considerations:

- locate the site within the same watershed, and
- replace on an acre for acre basis.

Also, it was agreed that the replacement wetland would be of the same system as the wetland to be filled. Therefore, only features for palustrine systems in this example will be evaluated.

Information summarized in Table 13 indicates that the three functions desired require consideration of 23 site selection features. This list of features can be reduced by deleting those which are not applicable to palustrine systems and those which are not applicable to the region. In this example, watershed size applies only to riverine systems and the vegetated canopy feature applies only to nontidal riverine systems.

Next, each of the site selection features should be applied to each potential wetland replacement site. This procedure allows for the comparison of many sites and also allows particular sites to be weighted by their ability to provide for a number of features relating to several functions.

It is important to note that each feature is also given a value for its importance to the function to which it applies. In selecting a site, an attempt should be made to choose a site which meets the criteria for as many features as possible. Because it is unlikely that one site will meet all of the necessary criteria, those features which have a high importance to function value should be considered first.

In this example, the features which carry a high importance to function include: soils and water alkalinity, water source, wetland/watershed ratio, diversity enhancement, regional diversity, special upland habitat features, sediment and contaminant sources, and human disturbance.

Having selected a site, the design features which will be incorporated into the wetland should be considered. Table 14 lists the design features which apply to the Nutrient Removal/Transformation, Wetland Dependent Bird Habitat Diversity, and Floodflow Alteration. There are a total of 25 site features which apply to palustrine systems in the region for the three functions.

Of the 25 features, only outlet characteristics are common to all three functions. The criteria and methods for each feature should be reviewed to determine if a conflict exists. The criteria for outlet characteristics for each function follows:

Section 3.6 *Outlet Characteristics*
Create a constricted surface water outlet or no outlet at all.

Section 6.2 *Outlet Characteristics*
Do not include a permanent outlet in the wetland. If an outlet is present it should be a constricted outlet.

Section 10.14 *Outlet Characteristics*
Create a permanent surface water outlet for the wetland.

Although each of these criteria is slightly different, a design could be developed which includes all three. Creating a permanent constricted outlet would meet the criteria of all three functions.

A second feature which is common to both Floodflow Alteration and Wetland Dependent Bird Habitat Diversity is water/vegetation proportions and interspersion. The criteria for each of the functions are as follows:

Section 6.6 *Water/Vegetation Proportions and Interspersions*
In headwater areas in particular, create a

wetland with a high proportion of vegetation in dense stands with little interspersed open water.

Section 10.3 *Water/Vegetation Proportions and Interspersions*
Create a well interspersed balance of open water and vegetation.

In general, when the criteria for features conflict, the feature which has the highest importance to function should be selected. In this example, Water/Vegetation Proportion and Interspersion for Wetland Dependent Bird Habitat Diversity (Section 10.3) would be selected.

All of the 25 design features must be considered in this way to determine how each can be incorporated into the design. Conflicts between criteria must be identified and resolved for each feature.

The end result of this site selection and site design evaluation will be a listing of the features which will make up a wetland replacement design.

References

Adamus, P. NSI Technology Services Corporation. U.S. Environmental Protection Agency Environmental Research Labs, Corvallis, OR. personal communication.

Adamus, P.R. 1983. *A Method for Wetland Functional Assessment: Volume II. FHWA Assessment Methods.* Federal Highway Administration FHWA-IP-82-24. p. 23-33.

Adamus, P.R. 1988. "Criteria for Created or Restored Wetlands". p. 369-372. *In:* D.D. Hook, W.H. McKee, Jr., H.K. Smith, J. Gregory, V.G. Burrell, Jr., M.R. DeVoe, R.E. Sojka, S. Gilbert, R. Banks, L.H. Stolzy, D. Brooks, T.D. Matthews, and T.H. Shear (eds.) *The Ecology and Management of Wetlands Volume 2: Management, Use and Value of Wetlands.* Croom Helm, London and Sydney.

Adamus, P.R., E.J. Clarain, Jr., R.D. Smith, and R.E. Young. 1987. *Wetland Evaluation Technique (WET), Volume II: Methodology.* U.S. Army Corps of Engineers Waterways Experiment Station, Vicksburg, MS. Operational Draft Technical Report Y-87-___ and Federal Highway Administration (FHWA-IP-88-029).

Ammon, D.C., W.D. Huber, and J.P. Heaney. 1981. "Wetland Use for Water Management in Florida". *ASCEJ* Water Resource Plan Management Division. 315 p.

Bennett, G.W. 1971. *Management of Artificial Lakes and Ponds.* Reinhold Publishing Company, New York, NY.

Blaney, Harry F. 1961. "Consumptive Use and Water Waste by Phreatophytes". Proceedings No. 2929, American Society of Civil Engineers.

Boesch, D.F. and R.E. Turner. 1984. "Dependence of Fishery Species on Salt Marshes: The Role of Food and Refuges". *Estuaries* 7:460-465.

Camfield, F.E. 1977. *Wind-Wave Propagation Over Flooded, Vegetated Land.* U.S. Army Corps of Engineers, Coast Engineering Research Center. Technical Paper No. 77-12.

Conservation Foundation. 1988. *Protecting America's Wetlands: An Action Agenda.* Final report of the National Wetlands Policy Forum. Washington, DC.

Cowardin, L.M., V. Carter, F.C. Golet, and E.T. LaRoe. 1979. *Classification of Wetlands and Deepwater Habitats of the United States.* U.S. Fish and Wildlife Service, Washington, DC, FWS/OBS-79/31.

Department of Natural Resources. 1967. *Guidelines for Management of Trout Stream Habitat in Wisconsin.* Madison, WI.

Federal Interagency Committee for Wetland Delineation. 1989. *Federal Manual for Identifying and Delineating Jurisdictional Wetlands.* U.S. Army Corps of Engineers, U.S. Environmental Protection Agency, U.S. Fish and Wildlife Service, and U.S.D.A. Soil Conservation Service, Washington, DC. Cooperative Technical Publication. 76 p. plus appendices.

Fisher, S.G. and A. Lavoy. 1972. "Differences in Littora Fauna Due to Hydrological Differences Below a Hydroelectric Dam". *J. Fishery Research Board of Canada* 29:1472-1476.

Fredrickson, L.H. and T.S. Taylor. 1982. *Management of Seasonally Flooded Impoundments for Wildlife.* U.S. Fish and Wildlife Service. Resource Publication 148. Washington, DC.

Fredrickson, L.H. and F.A. Reid. 1986. "Wetland Riparian Habitats: A Nongame Management Overview". p. 59-96. *In:* J.P. Hale, L.B. Best, and R.L. Clawson (eds.) *Management of Nongame Wildlife: A Developing Art.* N. Cen. Sec. Wildl. Soc., Chelsea, MI. 171 p.

Garbisch, Edgar W. 1986. *Highways and Wetlands; Compensating Wetlands Losses.* U.S. Department of Transportation, Federal Highway Administration.

Geological Survey Circular. 1971. *Real Estate Lakes 601-G.*

Knutson, P.L., J.C. Ford, M.R. Inskeep, and J. Oyler. 1981. "National Survey of Planted Salt Marshes (Vegetative stabilization and wave stress)". *Wetlands* 3:129-153.

Kusler, J.A. and M.E. Kentula. 1989. *Wetland Creation and Restoration: The Status of the Science.* Vol. 1. Environmental Research Laboratory, Corvallis, OR.

Kusler, J.A. and M.E. Kentula. 1989. *Wetland Creation and Restoration: The Status of Science.* Vol. 2. Environmental Research Laboratory, Corvallis, OR.

Marzolf, G.R. 1978. *The Potential Effects of Clearing and Snagging on Stream Ecosystems.* U.S. Fish and Wildlife Service FWS/OBS-78/14:32.

Meehan, W.R. (Technical Editor). 1982. *Influence of Forest and Rangeland Management on Anadromous Fish Habitat in Western North America.* Pacific Northwest Forest and Range Experiment Station. Portland, OR.

Mitsch, W.J. and J.G. Gosselink. 1986. *Wetlands.* Van Nostrand Reinhold Company, Inc., New York, NY.

Novitzki, R.P. 1979. "The Hydrologic Characteristics of Wisconsin Wetlands and Their Influence on Floods, Streamflow, and Sedimentation". p. 377-388. *In:* P.E. Greeson, J.R. Clark, and J.E. Clark (eds.) *Wetland Functions and Values: The State of Our Understanding.* American Water Resource Association, Minneapolis, MI.

Ogawa H. and J.W. Male. 1983. *The Flood Mitigation Potential of Inland Wetlands.* University of Massachusetts Water Resource Research Center, Amherst, MA. 164p.

Ploskey, G.R., J.M. Nestler, and L.R. Aggus. 1984. *Effects of Water Levels and Hydrology on Fisheries in Hydropower Storage, Hydropower Mainstream, and Flood Control Reservoirs.* U.S. Fish and Wildlife Service, U.S. Department of the Interior, and the Environmental Laboratory for the U.S. Army Engineer Waterways Experiment Station, Vicksburg, MS. Technical Report E:84-88.

Richardson, C. 1985. "Mechanisms Controlling Phosphorous Retention Capacity in Freshwater Wetlands". *Science* 228:1424-1428.

Seibert, R. 1968. "Importance of Natural Vegetation for the Protection of the Banks of Streams, Rivers, and Canals". *Freshwater Nature and Environment.* Series 2:35-67. Council of Europe.

Smith, B.D., P.S. Maitland, M.R. Young, and J. Carr. 1981. "Ecology of Scotland's Largest Lochs: Lomand, Awe, Ness, Morar, and Shiel, 7 Littoral Zoobenthos". *Monographs of Biology* 44:155-204.

Sousa, P.J. 1987. *Habitat Management Models for Selected Wildlife Management Practices in the Northern Great Plains.* REC-ERC-87-11. U.S. Department of the Interior, Bureau of Reclamation, Office of Environmental Service Technical Report, Engineering and Research Center.

Swales, S. and K. O'Hara. 1980. "Instream Habitat Improvement Devices and Their Use in Freshwater Fisheries Management". *Journal of Environmental Management.* Vol. 10. p. 167-179.

Turner, R.E. 1977. "Intertidal Vegetation and Commercial Yields of Panaeid Shrimp". *Trans. Am. Fish. Soc.* 106:411-416.

U.S. Department of Agriculture, Forest Service. 1985. *Fish Habitat Improvement Handbook.* Technical Publication R8-TP7. Forest Service Southern Region.

U.S. Department of Agriculture Soil Conservation Service. 1970. *Irrigation Water Requirements.* Technical Release No. 21.

U.S. Department of Agriculture Soil Conservation Service. 1971a. *Building a Pond.* Farmers Bulletin 2256.

U.S. Department of Agriculture Soil Conservation Service. 1971b. *Ponds for Water Supply and Recreation.* Agriculture Handbook 387.

U.S. Department of Agricultural Soil Conservation Service. 1971c. *Ponds: Planning, Design, and Construction.* Agriculture Handbook 590.

U.S. Department of Agriculture Soil Conservation Service. 1975. *Urban Hydrology for Small Watersheds.* Technical Release No. 55.

U.S. Department of Agriculture Soil Conservation Service. 1977. *Conservation Plants for the Northeast.* U.S. Government Printing Office, Washington, DC.

U.S. Department of Agriculture Soil Conservation Service. 1985. *National Engineering Handbook, Hydrology* 4: SCS/ENG/NEH-4-2.

U.S. Fish and Wildlife Service. 1988. *Waterfowl Management Handbook.* Fish and Wildlife Leaflet 13. U.S. Department of the Interior, Fish and Wildlife Service, Washington, DC.

Zimmerman, D.W. and R.W. Bachman. 1978. "Channelization and Invertebrate Drift in Some Iowa Streams". *Water Resources Bulletin.* p. 14868-14883.

Appendix A
Glossary

Ammonium volatilization - Abiotic process which results in the removal of ammonium by evaporation.

Annual flood - The highest peak discharge in a water year.

Annual high water - The highest water elevation from the annual flood.

Artificially flooded - The amount and duration of flooding is controlled by means of pumps or siphons in combination with dikes and dams.

Channel - An open conduit which periodically or continuously contains moving water.

Channel flow - Observable movement of surface water in a confined zone.

Class - The taxonomic unit used in the FWS wetland classification system (Cowardin et al., 1979) that describes the general appearance of the habitat in terms of dominant vegetation or some other feature.

Cluster - Wetlands situated so that there is a large number of wetland acres per total square miles. Clustered wetlands are not necessarily contiguous.

Constricted outlet - A surface outlet on a channel less than one-third the maximum width of the wetland, or a surface outlet on a standing body of water less than one-tenth the width of the wetland.

Deep water - Surface water depth greater than 6.5 feet and lacking vegetation.

Denitrification - The conversion of nitrate to gaseous nitrogen by microbes in anaerobic conditions.

Extreme high water of spring tides - The highest tide occurring during a lunar month, usually near the new or full moon.

Extreme low water of spring tides - The lowest tide occurring during a lunar month, usually near the new or full moon.

Fetch - The maximum open water distance unimpeded by intersecting islands, erect vegetation, or other obstructions.

Fringe wetland - Fringe wetlands along a channel. The total width of these wetlands is less than three times the width of the adjacent channel. Fringe wetlands on a standing body of water occupy less than one-third the surface area of a standing body of water at the time of highest annual water.

Functions - The physical, chemical, and biological process or attributes of a wetland.

Hydroperiod - The seasonal occurrence of flooding and/or saturated soil conditions.

Intermittently exposed - Surface water is present throughout the year except in years of extreme drought.

Interspersion - The degree of intermingling of different cover types, regardless of the number of types or their relative proportions.

Irregularly exposed - The land surface is exposed by tides less often than daily.

Irregularly flooded - The tidal water floods the land surface less often than daily.

Island - An area of land that is, at least, seasonally exposed, not connected to the mainland by any bridge, and uninhabited by humans.

Mean high water - The average height of high water in tidal waters over 19 years.

Mean low water - The average height of low water in tidal waters over 19 years.

Nitrogen fixation - The conversion of gaseous nitrogen into inorganic nitrogen forms by bacteria and blue-green algae.

Normal water table level - The average water table elevation during the growing season.

Oasis - One or more wetlands situated so that they are isolated or comprise a small number of acres per total square miles.

Overland flow - Runoff water originating from a rainstorm or snow melt which flows over the ground surface.

Peak annual flow - Peak discharge of the annual flood.

Permanently flooded - Water covers the land surface throughout the year in all years.

Piezometric surface - Imaginary surface to which water rises in wells in a confined aquifer.

Regularly flooded - The tidal water alternately floods and exposes the land surface at least once daily.

Saturated - The substrate is saturated to the surface for extended periods of time during the growing season, but surface water is seldom present.

Seasonally flooded - Surface water is present for extended periods especially early in the growing season, but is absent at the end of the season in most years. When surface water is absent, the water table is often near the land surface.

Semipermanently flooded - Surface water persists throughout the growing season in most years. When surface water is absent, the water table is usually at or very near the land surface.

Shallow water - Surface water depth less than 2 feet. Area usually vegetated.

Sheetflow - Water within a wetland which is not confined to a channel.

Spring tide - The highest high and lowest low tides during the lunar month.

Subclass - Subdivision of a class as used in the Cowardin et al. (1979) wetland classification system. Classes are based on substrate material and flooding regime, or on vegetative life form.

Subtidal - The substrate is permanently flooded with tidal water.

Surface water - Any water, temporary or permanent, above the ground surface, observable with an unaided eye.

Temporarily flooded - Surface water is present during brief periods during the growing season, but the water table usually lies well below the soil surface for most of the season.

Toxicant - Any substance present in water, wastewater, or runoff that may kill fish or other aquatic life, or could be harmful to the public health.

Unconstricted outlet - A surface outlet on a channel greater than one-third the maximum width of the AA or larger adjoining AA, or a surface outlet on a standing body of water greater than one-tenth the width of the AA or larger adjoining AA.

Urban Area - An area having a residential density of at least 1,000 residences/square mile over 4 contiguous square miles, or a central city having a population of 50,000 or more, and including surrounding, closely settled areas if these surrounding areas are (a) incorporated places of 2,500 inhabitants or more, or (b) incorporated places with fewer than 2,500 persons, provided that each place has a closely settled area of 100 permanent residences or more, or (c) small land parcels normally less than one square mile in area, having a population density of 1,000 inhabitants or more per square mile, or (d) other similar small areas in an unincorporated territory with lower population density when these areas serve to complete urban-suburban community boundaries.

Values - Wetland processes or attributes which are beneficial to society.

Water year - The year taken as beginning October 1. Often used for convenience in streamflow work, since in many areas streamflow is at its lowest at that time. Used by U.S.G.S. in their WSP.

Watershed - The upslope area from which surface waters enter the wetland replacement area and hydrologically connected wetlands at least seasonally. The watershed is measured from the replacement wetland's outlet.

Wetland-dependent - A term for species that may use nonwetland habitats, but occur in wetlands a preponderance of the year, or which have critical life requirements met by wetlands that are not provided by nonwetlands.

Wetlands - Those areas that are inundated or saturated by surface or ground water at a frequency and duration sufficient to support, and that under normal circumstances do support, a prevalence of vegetation typically adapted for life in saturated soil conditions. Wetlands generally include swamps, marshes, bogs, and similar areas (40 C.F.R. 230.3 and 33 C.F.R. 328.3).

Appendix B
Wetland and Native Plant Suppliers*

1. Country Wetlands Nursery, LTD, Box 126, Muskego, WI 53150
2. Environmental Concern Inc., P.O. Box P, St. Michaels, MD 21663
3. Fern Hill Farm, Route 3, Box 305, Greenville, AL 36037
4. Gardens of the Blue Ridge, P.O. Box 10, Pineola, NC 28662
5. Horizon Seed Co., 1540 Cornhusker Highway, Lincoln, NB 68500
6. Horticultural Systems, Inc., P.O. Box 70, Parrish, FL 33564
7. Kesters Wild Game Food Nurseries, Inc., P.O. Box V, Omro, WI 54963
8. Lilypons Water Gardens, Lilypons, MD 21717
9. Mangelsdorf Seed Co., P.O. Box 327, St. Louis, MO 63166
10. Mangrove Systems, Inc., 504 S. Brevard Avenue, Tampa, FL 33606
11. The Theodore Payne Foundation, 10459 Tuxford Street, Sun Valley, CA 91352
12. Plants of the Wild, Box 855, Tekoa, WA 99033
13. San Francisco Bay Marine Research Center, 8 Middle Road, Lafayette, CA 94549
14. F.W. Schumacher Co., Inc., 36 Spring Hill Road, Sandwich, MA 02563-1023
15. Sharp Bros. Seed Co., Healy, KS 67850
16. Siskiyou Rare Plant Nursery, 2825 Cummings Road, Medford, OR 97501
17. Slocum Water Gardens, 1101 Cypress Gardens Road, Winter Haven, FL 33880
18. Southern States Coop., 6606 West Broad, P.O. Box 26234, Richmond, VA
19. Stanford Seed Co., 809 N. Bethlehem Pike, Spring House, PA 19477

20. Stock Seed Farms, Inc., R.R. Box 112, Murdock, NB 68407

21. Wm. Tricker, Inc., 74 Allendale Avenue, Saddle River, NJ 07458

22. Van Ness Water Gardens, 2460 Euclid Avenue, Upland, CA 91786

23. Wildlife Nurseries, P.O. Box 2724, Oshkosh, WI 54901

* The listed sources do not include all existing ones; refer to local Soil Conservation Districts, U.S. Fish and Wildlife, Forest Service, native plant societies, arboretums, etc.